7

Mathsemantics

MAKING NUMBERS TALK SENSE

Viking

VIKING
Published by the Penguin Group
Penguin Books USA Inc., 375 Hudson Street,
New York, New York 10014, U.S.A.
Penguin Books Ltd, 27 Wrights Lane,
London W8 5TZ, England
Penguin Books Australia Ltd, Ringwood,
Victoria, Australia
Penguin Books Canada Ltd, 10 Alcorn Avenue,
Toronto, Ontario, Canada M4V 3B2
Penguin Books (N.Z.) Ltd, 182–190 Wairau Road,
Auckland 10, New Zealand

Penguin Books Ltd, Registered Offices:
Harmondsworth, Middlesex, England

First published in 1994 by Viking Penguin,
a division of Penguin Books USA Inc.

1 3 5 7 9 10 8 6 4 2

Grateful acknowledgment is made for permission to reprint excerpts from the following copyrighted works:
"Women in Statistics: Sequicentennial Activities," *The American Statistician*, May 1990, vol. 44, no. 2. By permission of *The American Statistician*. Copyright 1990 by the American Statistical Association.
"What Do You Care What Other People Think?" Further Adventures of a Curious Character by Richard P. Feynman, as told to Ralph Leighton. Reprinted by permission of W. W. Norton & Company, Inc., and Melanie Jackson Agency. Copyright © 1988 by Gweneth Feynman and Ralph Leighton.
Young Children Reinvent Arithmetic: Implications of Piaget's Theory by Constance K. Kamii with Georgia DeClark. New York: Teachers College Press. © 1985 by Teachers College, Columbia University. All rights reserved. Reprinted by permission of the publisher.
The Child's Conception of Time by Jean Piaget. English translation copyright © Routledge & Kegan Paul 1969. Reprinted by permission of Basic Books, a division of HarperCollins Publishers, Inc., and Routledge & Kegan Paul Ltd.
The Child's Conception of the World by Jean Piaget. By permission of Littlefield Adams Quality Paperbacks and Routledge & Kegan Paul.

ISBN 0-670-85390-9
CIP data available

Printed in the United States of America
Set in Postscript Century O.S.
Designed by Kathryn Parise

To my brother, Richard Henri MacNeal,
a better mathematician than I,
and in loving memory of
our mother,
Marguerite Josephine Marie Giroud MacNeal Cummin,
our father, Kenneth Forsythe MacNeal, and our sister,
Marguerite Louise MacNeal Lippincott,
to whom apples and oranges were
both a table decoration
and dessert

Acknowledgments

Quite unawares, Priscilla MacNeal, my wife, began work on this book in 1969 when she started trying to recruit people for our consulting firm who were "good at numbers." Two decades later, with her help and the directions my friend Edward D. Wohlmuth gives in *The Overnight Guide to Public Speaking*, I turned our recruiting experiences into a talk. (It later became the basis for chapter 1.) Gerard I. Nierenberg, of negotiating fame, liked the talk. To build on it, he introduced me to literary agent Robert J. Markel, who has shepherded this book ever since.

Along the way, Robert D. Loomis, Susan C. Milius, Robert S. Bramson, Stuart A. Mayper, Leslie Meredith, Richard H. MacNeal, Michael J. Brown, Catherine Wright MacNeal, Madeleine Schroeder, Edward F. Gardner, Richard P. Taylor, Thomas P. Crolius, Kendall Crolius, Jeffrey Mordkowitz, and Vincent Benedict commented on drafts, supplied materials, or both.

Editor Dawn Ann Drzal, copyeditor Theodora Rosenbaum, and others at Viking brought the book at last into print.

I am fortunate indeed to have been helped by so many truly talented and generous people. I give my deep thanks to each.

Contents

APPENDICES

CHAPTER 1

A fruitful start

If I were to start by saying that I had a *few brief* and preliminary points to make, your experience with such indefinite terms as "few" and "brief" might just cause you a little uneasiness.

When I say instead that I'm going to introduce *two* fields of inquiry and then combine them in *one* example, you probably sense that I know where I'm going and won't waste your time. Using even the two simplest numbers can make that difference.

The first field I have in mind is math. Call it numbers, figures, mathematics, abstract relations, or what you will. The second field is semantics. Think of it as the full range of meanings of any language or symbol-system, including math, working away inside us.

My own semantic interest in numbers stems from parents who spoke different number languages, a CPA father and a chanteuse mother.

I can still remember Mother murmuring her sums, usually bridge scores or prices scribbled on brown grocery bags. *"Trente, trente et soixante font quatre-vingt-dix, quatre-vingt-dix et cent font. . . ."* Most everything else except singing she did in English, but she said she never learned to add or subtract except aloud and in French. Dad, on the contrary, added only silently, very fast.

1

I also remember Mother's beautiful numerals, tall, thin, gracefully curved; her one with the long, almost vanishing, upswing preceding its more definite downswing, her seven with its honestly European cross-bar, and her nine gently curving down below the line. Compared with these beauties, Dad's numerals were just workaday dwarfs, cramped, not more than one-fourth the height or breadth of Mother's, and individually jagged even though lined up in precise rows.

Dad had exact ideas about numbers where Mother tended to round them out. I can remember her saying once that he was difficult when it came to money. If she owed him $1.02, she said, he insisted on getting the last two pennies; he wouldn't just take the dollar; and if he owed her 97 cents, he insisted on getting his three cents change. No wonder I learned, in the practical way children do, that when I needed to borrow money, ask Mother, but when I needed someone to hold funds for me, ask Dad.

My adult interest in numbers sprang from a career in market research and peaked sharply during two hundred cross-examinations under oath as an expert witness defending aviation market-research figures developed by my staff.

To survive the witness stand I needed to avoid error. To this end I devised in 1969 a recruitment quiz to test how well applicants avoided the kinds of numerical mistakes, mostly semantic in origin, I had by then made and noticed.

What I'll tell you in this book derives mainly from my own errors and the answers given by the one hundred ninety-six clerical applicants from 1969 through 1984 who had sounded the most promising of those answering our ads containing the qualifier "Good at numbers."

Apart from linguistic specialists, most people's acquaintance with semantics stems directly or indirectly from the general semantics of Alfred Korzybski (1879–1950), which enjoyed its greatest vogue in the 1940s. I remember Dad talking about Stuart Chase's 1938 popularization, *The Tyranny of Words*, when I would have been about thirteen, so I heard early on that the very language we speak might have some strange power to rule us.

General semantics certainly had then and still has profound implications. Many of these implications, however, were diluted as general semantics entered our culture's main stream. For example, Korzybski's dictum, "the word is not the thing," seems to have become "it's only a semantic problem," rather than the deeper puzzler, "whatever you say it is, it isn't."

As a practicing general semanticist, I have often derived a business advantage from semantic principles and their overlooked implications. Take the principle that cow_1 is not cow_2. No real cow is the same as the word "cow" nor is any real cow the same as any other real cow. $Thing_1$ is not the same as $thing_2$. Only our calling them by the same word makes it seem so.

It then follows that whenever we add two things together we add two *different* things. Any real-world addition, so to speak, adds apples and oranges.

If my third-grade teacher, Miss Chapman, was literally right that "we can't add apples and oranges," then it follows that only numbers can be added, not things. That's logical but impractical and operationally false. We do demonstrably add all sorts of "things" together. So what Miss Chapman must have meant was that the things we add must bear the same name. If we call cow_1 a "cow" and cow_2 a "cow," then we can have two cows. If we call $fruit_1$ a "fruit" and $fruit_2$ a "fruit," then we can add them even if $fruit_1$ is an "apple" and $fruit_2$ is an "orange." But Miss Chapman didn't explain this.

Business data depend on such naming abilities. To obtain company size we add clerks and managers together as "employees." To get a machine's full cost we add its price to its maintenance and other factors. In my business I regularly add airline scheduling, air fares, and airport accessibility together to obtain the datum we call a "service-value index," a useful number reflecting the convenience of making a round trip in any given city-pair market.

To discover which applicants had this number-related naming ability, the third question in my quiz asked applicants to solve the following combined-math-and-semantics problem in addition:

2 apples
5 oranges
_____

   The one hundred ninety-six applicants answering this question gave
fifty-six digitally different answers. I say "digitally," because if penman-
ship mattered, they gave one hundred ninety-six different answers.
Now, if I had to honor the uniqueness of each answer, I'd not be able
to discuss them with you in an intelligible manner. So for your convenience I've added the apples and oranges of the answers together in
seven groupings. I trust I have your blessing in this.
   The most frequent answer was the one I'd hoped for: "7 fruit." That
includes such answers as "seven fruits," "seven pieces of fruit," and
other variants. Fifty-two applicants gave this answer, or just over one in
four of all applicants.
   The next most frequent answer was "7." Just seven, not seven of
anything. Forty-six gave this answer, or just under one in four of all
applicants.
   The next most frequent answer was "2 apples + 5 oranges." I must
confess I fail to see how this solves any problem in addition. Thirty-six
applicants, or just over one in six, took this tack.
   More forthright, perhaps, were the applicants who gave the answers
I group under the heading, "can't add unlike values." I put twenty-seven
applicants, or more than one in eight, in this group, twenty-one because
they answered surrounding problems but skipped this one, and six because their answers explicitly rejected the addition of unlikes.
   A still less popular answer was "7 apples and oranges." That's not
bad. But without hyphens to make "apples-and-oranges" a new unit—
and they were never used—I don't see it as solving the naming problem in addition. Fifteen gave this answer, or about one applicant in
thirteen.
   Another fifteen applicants waffled. One wrote, "7 or 2 apple 5 oranges," which misses a plural, errs on both sides of the straddle, and
equivocates. You might say I was lenient to count such a reply as only
one error.

Finally, five applicants, or about one in forty, gave clear-cut wrong answers like "7 oranges." I doubt even today's genetic engineers could make two apples and five oranges into seven oranges.

Note that the answers resemble the party vote in some European elections. Each attracts a minority. The most frequent answer, "7 fruit," is the winner. The other answers don't solve the naming problem. Mostly they avoid it, by dropping all mention of units, by repeating "apples and oranges," by declaring a solution to be impossible, or by equivocating.

You can imagine how applicants protested this question. Lucky you. My wife, who's our personnel director, and I had to face the protests. I tend to side with the underdog, so I heard out the claims of unfairness and trickery. I agree that the protesters were duped. They were duped by a rule, "you can't add apples and oranges," that glosses over a very real semantic problem.

Outside math class we never add anything except "apples" and "oranges," so—and here's the semantic problem—how do we do it? No applicant I interviewed recalled having been told how. All recognized the "apples-oranges" problem, but none had developed a general solution. Indeed, none seemed ever to have addressed the semantic question, "If no two things are alike, what are the rules for adding things?"

Many applicants said they could have solved the problem had they known what answer we wanted. Think about that for a moment. It's either a particularly simple-minded statement or they're trying to tell us that they're willing to accept whatever the house rule is. Forget math and semantics.

Some applicants insisted the problem truly was not solvable. I'd then put my finger to my lips to signal a Korzybskian, let's-get-beyond-words silence, and, pointing to a red pencil and a green pencil, quietly say, "Please add these." Back came the answer, "Two pencils." "But," I lied, "in this office we call this [pointing to the green pencil] a 'greenie' and this [pointing to the red] a 'reddie.' You can't add greenies and reddies." "Oh yes I can," came the reply, "because they're both pencils."

"Oh," said I, adding a pen to the collection, "then what about

these?" "Three writing implements," I was now told without hesitation. A stapler made it "four desk implements," and my wastepaper
basket made it "five office objects."

"So," I chortled, "whether you can add things together depends on
what they're called, is that right?" This evoked only puzzled looks, so
I went on, "If I call them 'fruit,' you can add them together, but if I call
them 'apples' and 'oranges,' you can't. That makes it a question of who
will do the naming, doesn't it?"

Some still insisted that you could never add apples and oranges no
matter what they were called. I'd like to think of that not as a considered opinion but as a payback for putting applicants on the spot. Okay,
I'm at fault there. But I fear a few really believed the addition was impossible. And almost nobody wanted to think about it. That seems to
be part of the price we pay for divorcing arithmetic and semantics. The
divorce gulls us into treating math problems as "just math problems"
and semantic problems as "just semantic problems."

Anyone who accepts that you can't add apples and oranges attempts
the near impossible, to live without adding. It would never work in
business. The accounting department has to add apples, oranges, and
pomegranates every day. How can they? Easy. They just call them assets and express them in dollars. Sales, engineering, production, personnel; they all do that sort of thing, don't they?

The question is not whether we can add different things, but how we
can add them in clear and useful ways. That gets us into meanings,
into semantics, with both feet.

By looking at math and semantic aspects together, we can develop
a better feel for each. The logical purity of mathematical processes
such as addition then shines forth brightly. So does the semantic risk
of applying the process to the world. We thus reach a fact-of-life truth:
that all worldly additions, all extensional additions, combine different
things, and there are no final rules on how to do it.

One useful but incomplete rule for combining different things is to
find the narrowest common category. For example, in most business
applications, clerks and managers would be best totaled as "employees," not "people."

But note the dangers. "Employees" includes more than clerks and managers, and it slights the "people" aspect. No term captures what really goes on. The map is not the territory. If we miss this semantic aspect and things go awry, we may wrongly end up distrusting mathematics rather than our semantic habits. To make our numbers tell, we need to command both their math and their semantics.

And that's what I address in this book, how we should handle the combination of math and semantics. As you will discover, this combination includes more than enough critically important territory to be recognized as a "field" in its own right. I call this new field "mathsemantics." The word "mathsemantics," by the way, happily contains every letter in "mathematics" in the same sequence, plus just the letters s and n. Mathsemantics. Something New, slightly nutty, sounds nice.

Please be assured that this book is not a fussy theoretical treatise. My interests in numbers and semantics are deep but practical, based on my business needs and experience. I'm concerned about everyday failures to use figures sensibly, not abstract doctrinal questions. Although my subjects are math and semantics, I don't intend to sweat blood over definitions that don't matter. If given the choice, I'd rather say something approximately right about something important than say something precisely right about something unimportant. That's the approach I'll take here.

So, if you have picked up this book looking for a mathematical theory of meaning, you've been misled. My interest lies in the other direction, what we might call mathematical meanings in action or the semantics of numbers.

Sadly, traditional schooling mostly fails to solve the semantic problems involved in applying numbers to events. Educators say our schooling yields adults who fear to quantify or who fumble when they do attempt to quantify, adults apparently convinced that math is a powerful mystery they're not privileged to know. That schooling doesn't put math and semantics together. Indeed, semantics, if it appears at all, unfortunately ends up looking less like a key to understanding than a trivial fuss over words.

Well, you won't need any special knowledge of math or semantics to read this book. If you followed the apples-oranges argument, you're plenty intelligent—and open-minded—enough to grasp what I have to say.

Now, a promise. I promise you I've put no useless puzzles in this book, nothing designed just to show what marvelous things math can do, how smart I am (I'm not that smart), or how foolish people can look who don't know the math tricks.

Puzzles and tricks tend to annoy slower but surer thinkers. I still remember my brother's resentful complaint about geometry, a subject he reached three years before me. He said he didn't like geometry; it was just a bunch of puzzles. His work that day at the blackboard had been criticized, he said, even though it contained no errors, because he'd taken *more steps* than Euclid had.

Well, *this* book won't put you down for taking your own route. Quite the contrary. To "get" mathsemantics, I believe you need to think autonomously. The field of mathsemantics, as we'll see, covers too much of life to be reduced to a bunch of fixed rules and procedures.

Mathematics does have power, elegance, mystery, and even great beauty. But this book concerns plainer stuff, like apples and oranges, using numbers more tellingly, and not making damn fools of ourselves.

Nevertheless, every now and again, mostly at the ends of chapters, I will summarize a main point in a form that respectfully salutes all mathematics and mathematicians. Thus, to generalize our apples-and-oranges finding,

> **Proposition 1**: Whenever we add *things*, we necessarily add *different* things, which we must then group under the same *name*.

CHAPTER 2

The challenge of togetherness

Perhaps the most consequential social rebuff I ever received was at the age of five. I approached three girls of my age and acquaintance and started to join them sitting on the grass. "Go away," said one, "girls don't play with boys." Looking around, I could see the verdict was unanimous. So I acceded to their separatist demand, more or less for the next ten years. And for those ten critical years I was deprived. I lived beside girls, my sister was a girl. Yet my personal memories yield no insights into how it feels to be a girl from five to fifteen, how the world appears from that point of view.

Was it my pain at this sexist rebuff that led me ultimately to reject categories as rulers over what may or may not be brought together? I don't know.

But anyway, before I came to delight in adding apples to oranges, I first had to suffer twelve years of normal schismatic instruction. At school, math was clearly one subject and English, the only place where semantic-like analyses seemed welcome, was clearly quite another. Like girls and boys, math and English existed side by side but in their separate realms, their distinctiveness more presumed than explained. Math was definite and precise, wasn't it? And English less so, more feely-interpretive?

I can understand how its aura of precision entices us to exalt math. Like many people, I once believed that if accuracy and truth existed anywhere, mathematics must be the place. There was just no arguing a mathematical result. I remember my twelfth-grade report-card shock when my 92 in trigonometry plunged to a 63. Here was my first flunk since chorus in ninth grade when my voice went unruly. I summoned the courage to speak to our math teacher. "Mr. MacCormick," I said, "you gave me a 63 in trigonometry." "Yes, my boy," he replied, "and you earned every point of it."

Hah! No wonder I learned to use extreme caution with numbers. Who wouldn't fear such exacting authority?

Imagine this fear augmented in my virgin outing as an expert witness by the oath, "to tell the truth, the whole truth, and nothing but the truth." I had to defend estimates I'd constructed to show Philadelphia's need for an all-cargo transatlantic air carrier. The incumbent American-flag carriers, Pan Am and TWA, strongly opposed my estimates. Their attorneys had had several weeks to find holes in my exhibits. The cross-examination became pure torture. I did my novice best. I was cautious. Boy, was I cautious. I asked to have questions repeated. I looked for trouble and ambiguity in every word. I answered as carefully and narrowly as I could. I noted exceptions and stressed the sense in which I was answering. After two nerve-racking hours a recess was called.

Semantics then came to my rescue. Actually, it was a kindly judge, who put me wise to the meaning of my number-testimony. I've never mentioned it publicly before, for fear he may have broken some rule or other to help me out. "You needn't," he said quietly as he passed by, "be so circumspect."

"Circumspect," a magnificent word. "Circum," *around*, as in circumnavigate or circumference. "Spect," *look*, as in spectacle or inspect. Thus, "look around," which has come to mean "careful to consider all circumstances and possible consequences," not just "cautious." How could a judge tell me not to be cautious? That's different.

My Merriam-Webster *Collegiate* dictionary puts the difference this way. "CAUTIOUS implies the exercise of forethought usually prompted by fear of probable or even of merely possible danger; CIRCUMSPECT suggests less fear and stresses the surveying of all possible conse- quences before acting or deciding." What a judge! What an antidote to mathematical perfectionism! "You needn't be so circumspect."

Gradually in my career on the witness stand I came to accept that philosophical accuracy is a will-o'-the-wisp. What judges need is rea- sonable care—including usable numbers—and shorter hearings. Per haps a hundred costly people that first day sat through my four hours on the stand. Lawyers said not to worry. Hadn't Pan Am and TWA staved off competition that much longer and wasn't everybody there getting paid?

I'm grateful it was a judge who tipped me off. A real authority figure, one to match even Mr. MacCormick. I'd taken my oath solemnly in swearing "to tell the truth, the whole truth, and nothing but the truth." Never mind that general semantics had already convinced me you can't possibly say *all* about anything. The formula still took me in. I tried to do the impossible.

I continued to take the oath seriously, but never after that day quite so literally. I consider myself honest and forthright, but I've never re- fused to take the oath. In Quaker school I'd learned of martyrs who'd refused on religious grounds to doff their hats in court. Even so, I've never presumed to confront a judge on the absurdity of swearing under penalty of perjury to do something impossible.

Arizona Pima County Superior Court Judge Lillian S. Fisher, writing in *Newsweek* about the rules of evidence, details case after case barring witnesses from telling the whole truth. In one, a remarried widow sues for damages resulting from her first husband's death. She testifies under her original married name. Mentioning remarriage or her new husband's support would result in a mistrial. So would mention of her first husband's life insurance policy, social security payments to her children, and payment of his medical bills by his health insurance.

She sues for damages as a woman deprived of her dead husband's affection and future life earnings. What the jury hears is neither ex- actly true (her name, for instance) nor is it the whole truth.

Judge Fisher allows that it wouldn't be right to change legal rules

just because they belie common sense. I agree. They might just have
something else going for them. "But," she concludes, "in the name of
justice, we should at least change the oath that suggests that the witness will be telling 'the truth, the whole truth, and nothing but the
truth.' "

Mathematicians—well, at least many statisticians—dread the court-
room. I know this from their articles and editorials telling each other
how to handle cross-examination and also from my own work preparing witnesses, including, yes, even some math teachers. I believe
that most of them, like me that first day, try too hard to meet purely
mathematical standards and too little to meet mathsemantic ones.

Facing students' questions can give a teacher a false sense of authority and finality. In class, the teacher can be judge, jury, and execu-
tioner. But in court, the judge is in charge, your questioner is probably
being paid to make you look foolish, and every word you utter goes
into the record. Aaarrrrgh.

Still, there are things to like about courtrooms. Real issues get
heard and settled, not for all time, but for the practical now. Even unpopular theories, if related to the issues, must be heard out and
weighed.

I like the last most. My aviation analyses have sometimes worked
better than my opponents' because I know that a passenger isn't a person but something a person does. Now, that's unpopular, almost un-
grammatical. Yet it's good theory. It explains how one person can be
several passengers a year, a useful mathsemantic truth. A passenger
isn't a person, but something a person from time to time chooses to be.
An analyst losing sight of this temporal distinction risks turning any
passenger study—no matter its mathematical refinements—into just
so much high-sounding gobbledygook.

The challenge we face is not simply to understand math. That's too
abstract, too separatist, too elitist. No. The challenge is to understand
ourselves, how we do or don't put meanings into, and draw meanings
from, numbers and mathematical relations. The challenge is personal,
highly personal.

This challenge of personal interpretation affects you whether you're an atomic scientist, a housewife, or both, whether you're a business manager or a civil-rights lawyer, a soldier or a sales rep. We all use numbers or have them used on us.

No matter how much math or semantic sophistication you happen to command, neither by itself is enough. To meet the mathsemantic challenge, you must muster some of both.

If we could talk face-to-face, it might be fun to address your specific interests and experiences, to compare them with mine. Given this book setting, however, the best I can do is try to make it easier for you to bridge the personal gap. For that, you need to know a few things about my background.

My primary area of expertness is air-service analysis. I've often been cross-examined on my use of various mathematical and statistical techniques. I am, however, definitely not a *general* expert in mathematics or statistics. My credentials are strong in general semantics and not bad in decision making. I've been published in all of these fields— air service, semantics, and decision making. I've worked in factories and office buildings, as blueprint trimmer, executive secretary, market researcher, company president, and management consultant. I know more than most people about administrative law, music, direct mail, and a few other things, but I'm not an expert in them.

The examples I use reflect my interests. You will, I hope, be able to link them to your interests. I've included many from recent newspaper and magazine reports. Others reflect my business experiences or the answers of applicants to our recruitment quiz. Some are intensely personal. Internalized math meanings are what this book is about. Hence the order of its presentation generally flows from more immediate personal concerns to longer-range social ones.

The examples in the next chapter ("Problems with names") sample current mathsemantic errors. The examples in the two succeeding chapters ("The magic of names" and "The magic undone") illustrate the root cause of the errors: childhood semantics embedded in language.

The next twelve chapters explore particular mathsemantic problems at the individual personal level. The remaining seven chapters present a panorama of mathsemantic problems on a global interpersonal scale.

I hope you like to explore. I'm interested in all aspects of mathsemantics: the math side and the semantics side, childhood beliefs and stages, errors of all types, education and math anxiety, linguistic and cultural differences, evolution and history, math notation and number-memory, games and sports, childhood exercises that develop mathsemantic savvy, physical science and its philosophy, money and jobs, politics and the media, business and the professions, population and the environment, estimating and accounting, punctuality and time frames, gender differences, statistics and surveys, the future, what we can do about it, and so on; you name it.

How you approach this book obviously depends on you, whether you happen to be a word-person or a number-person or neither or both, your interests, your roles in life, and many other factors. I have no way of knowing, for example, whether you'd like to test your own mathsemantic savvy. If you would, you could tackle appendix A now or just answer each recruitment quiz question as it comes up. If, however, you'd prefer to skip self-testing in favor of other aspects of mathsemantics, no problem; just read away. It's entirely up to you. I believe in personal autonomy.

I do, however, have two requests: First, please relax your semantic expectations. Try to subdue any feelings you might have that you need a concise, exact, complete, and compelling definition of "mathsemantics" up front. Just let your feel for the field develop naturally as we go along. This isn't math class, and "mathsemantics" isn't in the dictionary yet. Even if you could find it there, dictionary words about words never quite give the whole sense of a new term, anyway, do they? Let the rough definition of mathsemantics (the combination of math and semantics) and the apples-oranges example serve as enough for starters. I'll have more to say on this point at the end of the chapter after the next.

My second request is also semantic: Don't fret if you don't understand everything in this book. The March 1992 *Notices of the American*

Mathematical Society reported that math teachers, feeling they *had* to "get it" when math researchers spoke,

> found it "enraging" that only five to ten minutes of the talks were comprehensible. Mathematicians commonly find themselves in this position during mathematics talks, but they usually don't worry about it.

So my second request amounts to asking you to read the way a mathematician would. Relax. *You don't have to get it all.*

Although mathematicians have written profound books on the foundations and meaning of mathematics, including such works as Whitehead and Russell's *Principia Mathematica* and Cassius Jackson Keyser's *Mathematics as a Culture Clue*, no one to my knowledge has written a whole book before on mathsemantics. The book you're reading is a first. It could give you an elegant, new, semantically oriented way of dealing with numbers and those who use numbers. It could even establish me as a new kind of expert.

Of course, on the down side, to put it delicately, it could also fail to establish the field. We'll see.

> **Proposition 2**: Although some people may qualify as experts, no one can say the whole truth about anything.

CHAPTER 3

Problems with names

Rollerskating home one afternoon at about age ten, I was set upon by two boys I'd never seen before. The bigger one—he might have been eleven—deftly dropped me to the street by the impressively simple act of stepping on the front of one skate. When I got up, he did it again. He clearly had the advantage. The smaller boy then called on him by name to put me down again. Casting about for any escape, I asked, "Is he your brother?"

"He used to be," said the older boy, "but he isn't anymore."

"Oh, I'm sorry," said I, grabbing the opening. "How did it happen?"

"Well," he replied, "we used to live together, but we don't anymore, because our parents split up."

We sat and talked, the three of us. They were pleased to hear how sure I was that they were still brothers, that their relationship didn't depend on living in the same house. I gave convincing examples, like brothers still being brothers after they grew up and left home. The fact that my own parents were separated gave us a bond. We parted friends.

A dozen years later I was struggling through Swiss psychologist Jean Piaget's *Judgment and Reasoning in the Child*. I was thinking how dense and open to doubt it was when I came upon a discussion of the term "brother." My pulse quickened.

What children mean by "brother," said Piaget, depends on their natural growth through three stages in understanding relationships. The first shows no recognition of relationship. Piaget quotes a six-year-old saying simply, "*A brother is a boy.*"

The second stage displays an intriguing mix of the first and third stages. A nine-year-old says a brother is, "*When there is a boy and another boy, where there are two of them.*" The investigator digs a little, "Has your father got a brother?" "*Yes.*" "Why?" "*Because he was born second.*" "Then what is a brother?" "*It is the second brother that comes.*" "Then the first is not a brother?" "*Oh no. The second brother that comes is called brother.*"

In the third stage the child achieves an adult's conception. "*A brother is a relation, one brother to another.*"

Because Piaget seemed right on the mark about brothers, I decided to test some of his findings that seemed to flow from improbable conversations. "Maybe French-Swiss kids think like that," I told myself, "but not American kids." So, heeding Piaget's warning that one must enter genuinely into a child's conversation, not just ask adult questions, I chatted with the children who would allow me.

Surprisingly I found five-year-olds who would talk with me about when brothers are brothers; about which came first, the city, the river, or the people living there; where the name of the moon is ("on the moon, of course"); and whether it would be okay if we all agreed to call the sun the moon and the moon the sun ("no, because the sun shines in the daytime and the moon at night").

Piaget was doubly right. Children don't think like adults and to discover that you must somehow enter the child's world.

What children say about names of things ("brother," "moon," "sun") suggests a psychological bond between word and thing reflecting a normal maturation of brain areas governing speech. Children naturally and unavoidably pass through a stage in which they feel words have power over things. Then using words correctly matters greatly.

To function as adults we need to retain some sense that the meanings of words should be respected. Unfortunately, our childhood stage
often imprints too strong a feeling that meanings have been settled by
custom and authorities. This is a problem; for good mathsemantics demands a challenging, investigative, and inventive attitude toward language. We must feel free to redefine apples and oranges as fruit,
missiles, paperweights, or theatrical props as suits our changing
needs. Then we can observe the rules of addition and have the result
make sense. But we must not let either semantic freedom or mathematical obligation dictate the entire tone.

The recruitment-quiz apples-and-oranges problem led immediately
to another on addition.

```
1 hr.   31 min.
2 hrs.  50 min.
6 hrs.  12 min.
```

The one hundred ninety-six applicants tackling this problem surprised us with eighty-four digitally different answers. Again, for convenience, I'll group them.

The most frequent answer was the one sought, "10 hrs. 33 min."
One hundred forty-three applicants gave this answer, or not quite
three out of four. We treated all other answers as incorrect.

Twenty, or about one in ten, gave answers that may have been mathematically right but were semantically wrong. Their defiance of idiom
obscured their meaning. The most frequent was "9 hrs. 93 min." My
favorites were "9 1/2 hrs 63 minutes" and "10 1/2 hrs 3 minutes."

"Hey, guy, meet me in one and a half hours and sixty-three minutes." What is this? Secret code? Fanny Brice doing her Baby Snooks
routine? No. Just a demonstration of semantically uncomfortable math.

Another twenty gave answers snubbing the semantic aspect entirely: "10   33," "10   13," and "9   93."

The remaining thirteen applicants, or about one in fifteen, displayed
the reverse problem. Good semantics, bad math. Examples: "10 hrs.

34 min.," "10 hrs 32 min.," both wrong by one minute, "10 hrs. 43 min.," "9 hrs. 55 min.," and "9 hrs. 33 min."

The next addition problem *specified the units for the answer*, but introduced a bigger semantic problem.

```
3 one-way trips
2 round trips
round trips
```

The most frequent answer was the one hoped for, "3 1/2 round trips." Eighty-four applicants gave this answer, or three in seven. They counted the three one-way trips as equivalent to one-and-a-half round trips.

The next most frequent answer was "2 round trips." Thirty-four gave this answer, or about one in six. Those I asked told me, "one-way trips can't be added to round trips." If they're right, then the majority is wrong. A few, probably remembering the apples-oranges ordeal, hastened to add that no one would ever change their minds.

Twenty-seven applicants, more than one in eight, answered "3 round trips." They converted two of the one-way trips to a round trip and disregarded the third.

Nineteen applicants, or about one in ten, answered "3 round trips & 1 one-way." We gave them credit, but conversations afterward suggested maybe we shouldn't have. "We gave you credit for this answer," I would say, "and you did a good job overall, but tell me, why didn't you treat the extra one-way trip as half a round trip?" "Oh, I could have done that," said some. "There just didn't seem to be room." They had me there.

But others answered, "The one-way trip may not be half a round trip; the person may not be coming back." "Then," said I, "if I have this straight, you're willing to assume that two one-way trips you know nothing about make a round trip but unwilling to assume that a third one-way trip is half a round trip. That seems inconsistent. How do you explain it?" They couldn't.

Sometimes I just asked, "How would you treat a one-way trip if you knew the passenger intended to return but did so in the next reporting period or died before getting the opportunity?" If the applicant came back with something like, "I see what you're getting at, intent's a poor basis for classification," that was a good sign. An applicant recognizing a bad move might still quickly pick up all the mathsemantics our business needed. Most applicants, however, just gave up.

The other thirty-two applicants, or about one in six, gave plainly wrong answers. In round trips, the highest answer was "8," the most frequent was "4 1/2," and the lowest was "1/2." Six didn't answer and three equivocated.

The term "trip" seems simple, doesn't it? Up close, however, we find that to make mathematical sense of counting air trips we must first solve a semantic problem. We must determine what we're counting.

I won't test your endurance with the details—really quite complex—of how the airline industry converts passenger itineraries into trips. The method isn't the point. Our unpreparedness to deal with mathsemantics, that's the point.

It struck some applicants as unfair when I asked them to decide whether a Boston–Seattle–Tokyo–Singapore–London–Boston trip, relentlessly westward, was a one-way or round trip. Or when I asked what the one-way trips were in a football scout's Philadelphia–Chicago–Green Bay–Chicago–San Francisco–Miami–Dallas–Philadelphia itinerary. There are no automatically right answers here. Our choices depend on what we expect to do with the "information."

Complex realities lie behind the term "trip." Counting "trips" involves semantic rules, arbitrary conventions, that aren't immediately clear. All the math in the world won't solve the semantic problem. But unless we do solve it, the appeal of precise math can easily lead us astray. Unfortunately, when we find we've been misled, we tend to doubt the math or the statistician. We tell ourselves, "Figures don't lie, but liars figure." We don't address mathsemantics.

Now, you may feel that these messy problems can be left to specialists. You might be right about air trips, but certainly wrong about the general principle.

Let's make it more personal.

Few people count air trips. Most use passenger counts. And they think they know what a passenger is; they've probably been one themselves lots of times. Yet the mathsemantic difficulties with "passengers" dwarf those of "trips."

Consider: *A passenger is not a person,* although almost always so regarded. A passenger is something a person does. This something varies by who does the counting and for what purpose.

As I reported in my 1981 specialized book, *The Semantics of Air Passenger Transportation,*

> In 1980 I was one passenger, ten passengers, eighteen passengers, thirty-six passengers, forty-two passengers, fifty-five passengers, seventy-two passengers, and ninety-four passengers. Each of these statements is true.

"How," you may well ask, "could this be?" The answer:

> I was one passenger in the sense that I was a person who traveled by air in that year. I was eighteen passengers in the sense that I made eighteen round trips. I was forty-two passengers in the sense that on forty-two different occasions I entered and exited the system of a different carrier. I was seventy-two passengers in the sense that on seventy-two occasions I was on board an aircraft when it took off from one place and landed at another. I was ninety-four passengers in the sense that I made ninety-four separate entrances and exits from airport terminal buildings.

Because these counts serve different purposes, all are used. Because we're unprepared for mathsemantic problems, we dismiss the potential confusion as a technicality.

> However ... a discrepancy in the meaning of *passenger* that allows my air travel ... to be counted alternatively as one pas-

senger, ninety-four passengers, and at various levels between, is too great to be ignored. Differences of this degree
are not merely technical. Unless we know more, we run the
risk of confusing one meaning of the word *passenger* with
another and projecting this confusion into our decision
making . . . about air transportation . . . possibly to our great
regret.

My book's specific arguments regarding air service and airport location may have had some effect. I can't say the same for its underlying
mathsemantic principles. That would have been too much to ask.
We've been conditioned all our lives to divorce math and semantics.

I'm not even claiming everybody in aviation got the word. They
didn't. NBC News on July 23, 1987, reported U.S. Secretary of Transportation Elizabeth Dole as saying that "last year 415 million people
were passengers on the airlines." Of course, that's absurd. Most of
these passengers were Americans, yet the population of the United
States in 1987 was only 244 million people, and about three-quarters of
them didn't fly at all that year.

The truth is there never were 415 million people, only 415 million
occasions when people boarded aircraft. Some, like me, were counted
fifty or more times. Almost everyone counted at all was counted at
least twice, once going out, once coming back. Some 65 million people
in 1987 were counted as 415 million airline passengers.

Politicians understandably want to emphasize that their activities
serve lots of people. Saying "415 million people were airline passengers" sways more votes than the complex truth. Elizabeth Dole is no
dummy. There's no law against mathsemantic pollution.

Aviation Daily reported on December 16, 1986, that Air Transport
Association vice-president Richard Lally, speaking about passenger security screenings, had said, "More than six billion people . . . have
been screened under the current program." The figure exceeded the
1986 world population.

At the local political level, Newark Mayor Sharpe James was quoted
in the February 1989 *OAG Frequent Flyer* as saying

> More than 23 million people [in 1987] came through Newark Airport. Their first, and often, last impression of our city was their contact with taxi drivers. That is why the image of these drivers is so important.

Good idea, perhaps, but bad mathsemantics. There weren't "23 million *people*," only 23 million *passengers*. At three round trips apiece, that boils down to about four million people. Perhaps half were North Jersey and New York residents, whose impression of Newark presumably had been formed long before. That leaves two million visitors. If half took taxis, then one million could have been subjected to taxi-driver contamination. What about mathsemantic pollution?

Journalists also tend to use the largest figures our mathsemantic ignorance permits. *Business Week* for December 19, 1988, in its article, "The Frenzied Skies," included under the title " MANY MORE PEOPLE ARE FLYING" a graph showing passenger growth from 250 million in 1978 to 440 million in 1988. Figures fitting the title would be more like 40 and 65 million. Not impressive enough?

The *Philadelphia Inquirer* for August 29, 1987, told us:

> During the course of 1986 alone, 31 percent of the adult population flew. In 1978, 275 million flew. This year, 450 million will.

Let's see now. If 450 million will fly, and they are only 31 percent of the adult population, then the U.S. adult population must be about 1.45 billion. No wonder census counts look understated.

In July 1987 the Airline Passengers of America invited me to join, offering this enticement:

> Last year over 415 million passengers enplaned on the nation's airlines. If they join forces, passengers can affect the policies of government and airlines.

That's a joining of forces I'd like to see: Me on a January 12 trip to Wisconsin joining me on a March 26 trip to San Antonio joining me on a November 20 trip to South Bend joining me on a December 22 trip to Los Angeles joining me on a December 29 trip to Philadelphia. A new kind of time travel!

Business Week for June 11, 1990, in its article "Taiwan's Dollar Offensive Is Gaining Ground in China," reported that "since 1987, more than 1 million Taiwanese have traveled to China." I don't know whether to be impressed. I can't stop to check all these things out. If the report is true, then about one in twenty Taiwanese has visited the mainland since 1987. That would show strong ties. If, however, the figure is a three-year *passenger* count, then any twenty passengers could be just one person making ten round trips; so perhaps only one in four hundred Taiwanese visited the mainland. That's a different story, isn't it?

William Zinsser, in his book *On Writing Well*, after filling eight pages with examples of bad English usage, continues: "I could go on. I have enough examples to fill a book." But he doesn't do so, and neither will I. However, I do want you to get a real taste of this particular mathsemantic offering, the habit of mistakenly counting events as people. So, here are the final aviation tidbits and then some other goodies, all leading inevitably to one conclusion.

America West's chairman, Edward R. Beauvais, in a prepared statement to the Aviation Subcommittee of the U.S. Senate Committee on Commerce, Science, and Transportation on November 8, 1989, said, "Growth has brought America West to the point where we now provide service to more than 13 million customers annually." "Customer" sounds like "people" but means "passenger." The number of different *people* America West serves in a year is probably about 2 million. When in Rome, do as the Romans do.

Six weeks later *Aviation Daily* reported that "Airlines serving Phoenix will be reimbursed [$900,000 for prorated costs] by America West because it has misrepresented annual passenger traffic since 1984." America West had defined "passengers" so differently from the other carriers that the 1988 counts alone differed by 3,253,402. In Phoenix, as elsewhere, a penny saved looks like a penny earned.

Boeing advertised in November 1989 that "by this time tomorrow, nearly two million people will fly on Boeing jetliners." Small type explained the claim counted every passenger boarding a Boeing jet during twenty-four hours. It failed to note that every passenger connecting from one Boeing flight to another would be counted again. Does Boeing, whose order backlog reached $84 billion in June 1990, need this hype?

Counting traffic as people also occurs on the ground. *Airport Magazine* reported in its January 1990 issue that

> reducing auto traffic is a main benefit of the 13-mile loop track system at D/FW [Dallas/Fort Worth] Airport. More than 8.5 million persons used this system last year, and by far the largest number—about seven million—were airport employees.

How'd you like to meet that payroll?

Aviation has no monopoly on counting repetitive actions as people. It's just the field in which I work. Examples keep crossing my desk. But enough aviation.

Human services: *Newsweek* on October 7, 1985, noted in its article "How Many Missing Kids?"

> At the center of the storm is the dispute over just how many children *are* missing. In 1983 the U.S. Department of Health and Human Services put the number at 1.5 million a year—a figure that has been widely circulated by private organizations. But experts say roughly 95% of those are runaways —many of whom return home within days and are counted repeatedly if they run away more than once a year. The vast majority of the rest are taken by estranged parents in custody battles. Only a tiny fraction fall into the category parents fear most—abductions by strangers—and even that figure is in dispute. Jay Howell, executive director of the National Center for Missing and Exploited Children, a nonprofit clearinghouse set up by the government in 1984, estimates that between 4,000 and 20,000 children are kidnapped by strangers

each year. [Bill] Treanor [executive director, American Youth
Work Center] says the figure is closer to 100, and the FBI
logged only 67 such abductions last year.

Kidnapping is a terrible crime. An uncertainty ranging from
1,500,000 to 67 is a huge uncertainty. That's 22,388 to 1.

Medical services: The *Philadelphia Inquirer*'s lead story of January
23, 1989, headlined "Venereal disease surges—as city scales back
clinic," prompts the mathsemantic question, "Is a patient a person or
a visit?"

In 1985, 25,588 patients were seen at the one city-run clinic
that is devoted solely to the treatment of sexually transmitted
diseases. Last year, the number of patient visits at the so-
called STD clinic was down to about 18,000—a 30 percent
drop-off from three years earlier.

Communication: AT&T ran an advertisement in March 1989 under
the headline, "For 65 million callers, every public phone is now an
AT&T phone." The small type read, "Sixty-five million people choose
AT&T each day." So "callers" equal "people." AT&T's computers
count telephone calls. Bad mathsemantics makes "calls" into "callers"
and "callers" into "people." (I ignore for our purposes here that
"choose" is a semantic corruption of "use.") If 65 million calls a day
use AT&T lines, they certainly are made by fewer than sixty-five million people. If duplication doesn't matter, why not multiply by 365 and
say that 24 billion (24,000,000,000) people "choose" AT&T each year?
Or does that make the absurdity of counting each telephone call as a
person too obvious?

Religion: *Business Week* for October 15, 1990, reports that "there are
officially 217 million followers of various religions" in Japan, "a country of 122 million." A follower, then, is not a person, but something a
person does.

Education: The *Smithsonian* for November 1984 carried an article
stating that McGuffey Readers have been

the most widely circulated textbooks in the United States. Approximately 122 million copies of the McGuffey Primer and the six graded Readers were sold between 1836 ... and 1920.... Perhaps a billion American children during these years imbibed the McGuffeys' brief moral tales.

If I wrote letters to editors regarding mathsemantic errors, I'd have no time left. Nevertheless, I lifted my ban and wrote the *Smithsonian*.

While *McGuffey Readers* were undoubtedly popular, they could not have been read by "a billion American children" ... because there haven't yet been a billion Americans in total, let alone that many school children between 1836 and 1920.... I estimate that only about 110 million different children attended school in the years 1836–1920.... I wonder whether each child for one school year was counted as another child. Adding annual school enrollments together would overstate the number of children.

An associate editor wrote politely thanking me for writing. "We cede to your unarguable mathematics. The authors based their figure on the supposition that perhaps ten children used each book, then multiplied by 122 million [the books sold]. We should have caught the fallacy in the checking process. Unfortunately, we did not."

Take heart, *Smithsonian*. When it comes to missing mathsemantic fallacies you're not alone.

A friend sent me a Bil Keane "Family Circus" cartoon. A little girl behind a lemonade stand is explaining to a little boy, "We've had five customers, and three of them were Daddy." Now, *that* I like. Super! Attagirl! Tell it like it is.

Proposition 3: For a count to make sense, you have to know what you're counting.

The magic of names

O nce, when Noël Coward was a young man, as I recall the tale, he was invited to dinner at one of those famous London gentlemen's clubs. It was to be a new experience for him, an important step into society. After arriving alone and being ushered in, he paused to survey the clubroom just below. The members were in street clothes. From inexperience he had dressed formally. Instantly, however, he announced to all, "Now, I don't want *anyone* to be embarrassed."

What reminds me of this story is something I've stumbled across while preparing this chapter. What I wanted to do was review for you some of Jean Piaget's findings regarding children before turning, in the following chapter, to some similar things semanticist Alfred Korzybski said about adults. Their findings have fundamental implications for mathsemantics and I wanted to get them right.

So I borrowed my wife's psychology textbook, a monster four-color volume in its tenth edition since 1937. Piaget's listing in the index directed me to eighteen different pages. Only Sigmund Freud's listing was longer.

I scanned the material quickly. It looked great. Page 220 credited Watson with behaviorism, Freud with psychodynamic theory, and

Piaget with the cognitive approach. Eighteen pages later was a photo-
graph and a full-page, two-column profile of Jean Piaget. It described
the great man as "one of the most prolific, if not the most famous psychologist of the century."

"Piaget," I read, "has only been widely known in this country since
the 1960s . . ." and I felt proud, like a true pioneer, for only that day I'd
located a term paper on Piaget I'd done at the University of Chicago in
1948. "Here you go," I complimented myself. "You spotted him a
dozen years ahead of the pack."

My eye completed scanning the sentence, ". . . when his works
were translated from their original French."

"Uh, oh," I thought, "something's wrong here. I never read French
that well." I consulted my 1948 paper. It listed five translations of
Piaget's books published in 1926, 1928, 1929, 1930, and 1948.

Somewhere I've read that nothing turns off a reader or viewer quite
so fast as discovering a clear material error. My wife just yesterday
called my attention in a TV movie to a plaster bust we'd seen broken—
tossed out a window—two scenes earlier. We never got back into the
flow of that epic.

As you can imagine, I read the other references to Piaget in my
wife's text with a pinch of skepticism. The further I went, the more embarrassed for the authors and publisher I became. I even began to
wonder if the authors and I had read the same Piaget. (No kidding; he
wrote so much that it's possible we consulted different volumes.) They
stressed Piaget's finding that children's cognitive abilities mature in
stages. They all but ignored the semantic contents of those stages, and
they annoyingly summarized his methods as "clever questions."

What had struck me in 1948 and still does today is how Piaget
brought home the point that to understand a child's thinking we must
judge it not as adult thinking full of errors but on its own terms as a different kind of thought.

Piaget repeatedly stressed the difficulty of entering the child's
world, of listening carefully to weed out convictions suggested by the
adult, the child's romancing, and answers at random. The last are what
Harvard psycholinguist Roger Brown has called, in honor of a child's

actual and irrelevant answer to an adult interrogator's question, "Pop goes the weasel."

From the little I did on my own to test Piaget's conclusions, I can attest to the difficulties; for fumble I did. But even my amateurish conversations with children convinced me, as I've already noted, that Piaget knew what he was talking about.

Unless you've tested them yourself with children five to ten years old, Piaget's conclusions may seem strange. That strangeness, however, has a bearing on mathsemantics, so let's take a look at what he said.

Each child, according to Piaget, has beliefs "never communicated to anyone," because the child regards his or her perspective as immediately objective and presumes everyone thinks the same way.

In the child's "reality" nothing is impossible and nothing obeys causal laws. Even the most fantastic events children believe in can be justified by motives. Children believe that the sun and mountains were made by people, that Lake Geneva, for example, was dug by workmen who then filled it with water for the people. Children also look upon fire, wind, sun, and moon as alive and conscious because they move by themselves.

Further, the child seeks a reason for everything. The idea of chance, of the accidental or arbitrary, is absent. Yet children's beliefs don't form a system, and children don't feel a necessity to avoid contradictions in their successive opinions. The child seeks understanding of the whole and does not engage in analysis of the parts, so that, for example, words not understood are allowed to slip by and what is said may be completely misunderstood. Finally, children confuse their thoughts with words and the words with the things they're thinking about. They identify thought with using words. They believe that thinking, silent thinking, is done "with the mouth."

Piaget filled volumes with a rich store of overlapping and interwoven examples of conversations by and with children in support of these points. These examples, especially if you will take the trouble to verify them yourself using Piagetian techniques, are persuasive.

I will here present examples from just one book, *The Child's Concep-*

tion of the World, and on just the one subject most immediately important for our purposes, the problem of names. Piaget regarded it as central. "The problem of names," he said, "probes to the very heart of the problem of thought, for to the child, to think means to speak. And if 'word' is a somewhat vague concept to the younger children . . . what is meant by a 'name' is on the contrary quite clear."

For Piaget, "questions on names" (for example, "What is that?") "characterize the most primitive stages of a child's questioning" and "in learning the names of things the child . . . thinks it is reaching to the essence of the thing and discovering a real explanation." For a child, learning a thing's name solves the problem of what that something is.

Although not obvious, this point really matters for mathsemantics, so let's explore it closely through examples.

First, children at age five or six know that a name is "to call something by," but treat it as a necessary attribute of the thing.

> "Why is the sun called what it is?" *Because it behaves as if it was the sun.*
> "Why are clouds called like that?" *Because they are all grey.*

Therefore, things could have only the names they have.

> "Could the sun have been given another name?" *No.* "Why not?" *Because it's nothing else but the sun, it couldn't have another name.*

And the name wasn't "given" to the thing, but came with it.

> "Where does the name of the clouds come from?" *The name? That is the name.* "Yes, but where does it come from?" *The clouds.* "What do you mean when you say it comes from the clouds?" *It's the name they've got.* "But how did the name happen? How did it begin?" *By itself.* "Yes, but where did the name come from?" *By itself.*

So, naturally, things couldn't have existed before they had names.

> "Before the sun had its name was it already there?" *Yes.*
> "What was it called?" *The sun.* "Yes, but before it was called
> sun was it there?" *No.*

And things that don't exist couldn't have names.

> "If a thing wasn't there, could it have a name?" *No.* "Long
> ago men used to believe there was a certain fish in the sea
> which they called a 'chimera' but there wasn't really any such
> fish ... so can't a thing that doesn't exist have a name?" *No,
> because when God saw that the things didn't exist he wouldn't
> have given them names.*

Whereas whatever has a name must exist.

> A little girl of 9 asked: "Daddy, is there really God?" The fa-
> ther answered that it wasn't very certain, to which the child
> retorted: "There must be really, because he has a name!"

Children readily combine these beliefs with their other beliefs, for
example, with their animism.

> "Does a fish know it is called a fish?" *Of course.* "Does the
> sun know its name?" *Yes, because it knows it's got its name.*

These beliefs resist being dislodged by argument.

> [To a child who has admitted that God might have changed
> the names of the sun and moon:] "Would they have been
> right then or wrong?" *Wrong.* "Why?" *Because the moon must
> be the moon and not the sun and the sun must be the sun!*

Bilingualism doesn't seem to help.

"Could this chair have been called 'Stuhl'?" *Yes, because it's a German word.* . . . "Have things got more than one name?" *Yes.* . . . "Could the sun have been called 'moon' and the moon 'sun'?" *No.* "Why not?" *Because the sun shines brighter than the moon.* . . . "Yes, but the sun isn't changed, only its name. Could it have been called . . . etc.?" *No.* . . . *Because the moon rises in the evening and the sun in the day.*

These views give way unsteadily to adult ones.

[To a child two months shy of eleven:] "Have words got strength?" *It depends on the word.* "Which ones have strength?" *The word 'boxing' . . . oh, no, they haven't any strength* (laughing). "Why did you think they had [at] first?" *I was wrong. I was thinking it was the word that hit.*

The problem, noted Piaget, "is more than verbal." It amounts to a systematic "confusion between the sign and the thing signified," even to the degree that "the child cannot distinguish a real house, for example, from the concept or mental image or name of the house." To a child, he said, "every object seems to possess a necessary and absolute name, that is to say, one which is a part of the object's very nature."

"It will, therefore," wrote Piaget, "be interesting to see at what age children can distinguish the word which designates it from the thing itself." His conclusion: Up to age seven or eight, children make no distinction between word and thing, they fail to understand the name-thing problem; then until about age eleven children understand the problem but can't solve it systematically.

My question: When, if ever, do adults really "solve it systematically"?

I remember from when I was about eight Mother's making some excuse for Dad's being late for dinner. It seemed he was seeing a sick friend, but her manner clearly indicated no questions were to be asked. I ultimately—and I mean *years* later—concluded that Dad had

been giving blood and that the friend died of a then unmentionable disease, which an army buddy of mine later called the "awful awful,"
which I still later heard referred to as the "big C," and which we now
openly call cancer. It isn't only the name of God, is it, that must not be
known (old belief), said aloud (later belief), or taken in vain (current
version)? We put Lenny Bruce in jail for saying things in clubs that you
can now hear on television. I hesitate to say them here, but you know
what I mean.

Verbal taboos, of course, are only a particularly public kind of word
magic. The strange beliefs of Piaget's children about motives, names,
and reality may accompany them in some fashion for the rest of their
lives. My wife's text says that Piaget assumed that "in each stage of
cognitive development the ways of thinking that characterized earlier
stages are not lost but rather remain available in 'higher' form, for application in suitable circumstances."

What might those circumstances be? How about in mathsemantics?
Let's try it.

Take apples and oranges. Vestiges of childhood beliefs could predis-
pose people to presume unconsciously that names tell the essence of
things, and that everything, including math, has its own motives.
Okay. Then they could well hear "you can't add apples and oranges"
as a mathematical law forbidding the combining of any two things
whose essences are different. The ban would seem to make sense. Apples can be added to apples, because their essences are the same. Apples can't be added to oranges, because their essences are different.
But hours and minutes can be added to hours and minutes, because
their names and essences are the same.

Adding one-way trips to round trips within this scheme, however, becomes problematical. Are their essences the same or not? Strictly on
their full names, their essences are different. However, I don't recall
applicants making that argument, only their nonplussed looks upon
my suggesting that their willingness to add depended on what things
were called. Instead, when I asked why they wouldn't convert one-way
trips to round trips, they told me directly that the two kinds of trips

were different. They then backed this up immediately with some example that had apparently decided the issue for them.

They seemed unaware of the "name" or semantic component. To them the question reduced to whether the things to be added were the same or different.

This also fits Piaget's description. Our applicants didn't distinguish their words from their thoughts or from what was really there (as they saw it).

People thinking this way, who don't distinguish names from things, could well objectify "passenger" as "a person in a conveyance." That fits the general scheme that nouns represent things. Adding passengers to passengers would raise no doubt, because their essences are the same. Regarding the sum as people would also raise no doubt. We thus arrive at an explanation of such absurdities as having more people in conveyances than there are people. Further, we can begin to understand why these absurdities seem so natural.

The same kind of reasoning applies to the errors reported earlier that arise from treating counts of "callers," "customers," "patients," or "students" as counts of people.

Now, reasoning in essences presumably given by names works part of the time. Two dollars plus five dollars makes seven dollars. And two oranges plus five trips probably wouldn't yield anything useful.

However, reasoning through essences makes us slaves to the essence namers, whoever they were. Like the express company agent in Ellis Butler's 1905 short story, we fall into the trap of reasoning that "pigs," even if they're guinea pigs, "is pigs." And one might add, "one-way trips is never round trips." Our recruitment quiz seems to have turned up some extreme cases of this approach.

Such reasoning also fails, as we've seen in the passenger example, when the name directs us to the wrong mental construct. Once again, however, no difficulty may arise in a narrowly constrained application. For example, treating eighty passengers on a plane as eighty *people* works smoothly within the cabin walls. It's when we add these eighty passengers to eighty on another plane, say, going in the opposite direction the next day, that we risk counting some people twice. The further we extend the addition, the more dramatically the reasoning fails. When we extend it to the whole country for one year, it fails colossally.

My firm avoided hiring people who reasoned like Piaget's charges. Our work extended beyond immediate perceptions and was tested on the witness stand. We got paid for doing it in a professional manner.

However, I did interview many applicants who had mangled the mathsemantic questions but otherwise had good qualifications. Some of them caught on quickly. They added red pencils to green pencils to pens to staplers to wastepaper baskets with gusto. Their eyes smiled as we cut the apples-oranges knot.

Others balked. They stuck stubbornly to the argument that you couldn't add apples to oranges, or one-way trips to round trips, no matter what you called them. They puzzled me and still do; for who can ever know another person's unspoken thoughts and methods.

Part of their stubbornness may have sprung from their annoyance with me. But that explanation may be too glib, too self-centered. What cause has any of us to dismiss what these applicants apparently sincerely believed? That thought should keep us open to other explanations.

Anyway, for me, with the benefit of hindsight and the kinds of insight covered by this chapter, I've begun to sense how these interesting applicants might have reached the answers they felt so sure were right.

> **Proposition 4:** Childhood semantics can impair adult mathsemantics.

Childhood semantics can impair not only mathsemantic particulars, it can affect our view of the whole field. You may recall that near the end of the second chapter I asked you to relax your semantic expectations, to forget about getting an immediately compelling definition of mathsemantics, and to let your feel for it develop as we went along.

You may have passed quickly over that request, because you normally delay your expectations when working with new things. If so,
great.

However, not everyone works that way. One of my draft readers, for
example, stressed repeatedly that I should "tell readers clearly what
mathsemantics is," in order "to fully establish and explain it," before
shifting to "our mathsemantic problems and what to do about them."

This is an utterly natural feeling. Piaget found that children believe
names are integral parts of things, that whatever exists must have a
name, and that whatever has a name must exist. Accordingly, my using
the name "mathsemantics" suggests there's some thing that corre-
sponds to that name. I understand the impulse.

Resisting the impulse, however, seems to promote sound mathsemantics. For example, as we've seen, there's no "thing" that corre-
sponds to the word "passenger," only repeatable actions. Immediately
defining "passenger" as, say, "a person in a conveyance" has to make
it harder, not easier, to discover the truth behind passenger counts.

The same is true of "mathsemantics." We need to play freely with
combining math and semantics, explore the problems that turn up,
keep our minds open, and not be afraid of having fun. We mustn't let
a definition get in our way.

If *forced* to hazard a definition, I'd probably go with something like:
"Mathsemantics is the science of the semantics of mathematics." I'm
not sure this will last. But even if it lasts, I hope it doesn't satisfy you.
It's flavorless, too abstract.

Let's face it. When we speak of mathsemantics, we can't point to any
thing, like a horse or a fireplug, with definite outlines. Instead, we must
mentally blend two fields, math and semantics, each a model of abstraction, formlessness, and change. No wonder we don't immediately
lock onto a crisp image.

Semantics is usually defined as "the study of meanings." Now, whatever interests people means something to them, so *everything* that interests people—including mathematics—has semantic aspects.

"Meaning," of course, is a muddy pool of undetermined depth. The child's views that thinking is with the mouth and that everyone thinks the same way seem straightforward enough. But adults see more complexity. A famous 1923 book on semantics, by C. K. Ogden and I. A. Richards, bore the title *The Meaning of Meaning*. George Lakoff took forty-six pages in *Women, Fire, and Dangerous Things* to elucidate the meanings of one common word, namely, "over," and one hundred twenty-four pages to handle "there" constructions ("there is," "there are," "there was," and so on). General semantics stresses physiological and psychological involvements that leave ordinary linguistic studies far behind. Well, when I use the term "semantics" I mean to use it in its broadest sense.

Mathematics continues to grow and expand its already vast realm. My favorite single work on math remains the 1956 four-volume best-seller, *The World of Mathematics*, which describes itself as "A small library of the literature of mathematics, from A'h-mosé the Scribe to Albert Einstein, presented with commentaries and notes by James R. Newman." Its one hundred forty-three selections from the originals plus Newman's commentaries give math a substance no definition or specialized text could ever achieve. Yet Newman's book has now unfortunately fallen behind the times; look in its index and you find no fractals, no fuzzy sets, and no computers as such, only "computing machines." Well, as with semantics, when I use the term "math" I mean to use it in its broadest sense, including all ordinary uses of numbers.

My efforts in mathsemantics do not, *could* not, exhaust the field. They merely point to examples and group them under a new *name*. I do not claim to have discovered mathsemantics, the *thing*. You'll find many quotations in these pages showing that others thought and wrote about math-and-meaning long before I did.

Yet what I've written here is original, created mostly from my own experiences and my own thinking, to the small extent that any of us ordinary people have the joy of going beyond what the great ones have taught us.

Someday, when the mathsemantics field has been repeatedly plowed and reaped, and perhaps left fallow, scholars may catalog its past produce in regimented textbooks. But right now the field is fresh and fertile. Why not then sample each fruit as a potential new delight?

> **Proposition 5:** Some things that can be neither counted nor measured can still be savored.

The magic undone

Nobody seems offended by the fact that children under eleven think that words and things are inseparable, that names are part of the essence of things.

Children don't read Piaget and presumably wouldn't be offended if they did. We adults can just regard the word-thing confusion as something we've outgrown, if we ever thought like that; for who remembers? Psychology textbooks, like my wife's, can bury Piaget's semantic discoveries under a blanket of labels (such as cognitive approach, developmentalist, nominal realism, cognitive stages, sensorimotor, preoperational, concrete operational) and abstruse explanations. The gripes of a few students required to regurgitate these abstractions on exams don't disturb the general tranquillity.

But then along comes Alfred Korzybski.

His 1933 magnum opus, *Science and Sanity: An Introduction to Non-Aristotelian Systems and General Semantics*, accuses *adults*, in effect, of the attitudes toward words that Piaget found in children. And whereas Piaget had been content simply to report what children think, Korzybski went on to tell adults what they *should* think.

Piaget said children don't distinguish the word from the thing.

Korzybski admonished, "The word is not the thing, the map is not the territory."

Piaget said children think they understand a thing when they know its name. Korzybski warned, "Whatever you say it is, it isn't."

Piaget said children believe words convey the essences of things. Korzybski lectured we can't know essences, only structures and relations.

Piaget said children eventually learn adult dichotomies between word and thing, thought and reality, word and thought. Korzybski said we must distinguish a sequence of three levels: event, object, and symbol.

By "event" he meant whatever whirligig of multidimensional forces is "out there," not directly knowable, only presumed to explain our sense impressions. We usually ignore these forces in favor of our constructed images at the next level.

By "object" he meant the nonverbal constructs in our heads given by our brain's particular way of processing our sense impressions. Humans seem mostly to use constructs based on vision, dogs go more for smell, bats for high-frequency sonar, rattlers for heat gradients, frogs for moving black dots, sharks for erratic vibration. These constructed images, our familiar objects, are not the world, Korzybski said, no matter how obvious they seem.

By "symbol" he meant the ordinary verbal level, but also mathematics, pictographs, and whatever we use to stand for something else.

Piaget said children's beliefs mature gradually and naturally into adult beliefs. Korzybski decried most adults' beliefs as immature.

Piaget labeled children's equating-their-thoughts-with-things "realism" and their equating-words-with-things "nominal realism." Korzybski called such equating "identification of different levels of abstraction" and condemned it.

Korzybski said human sanity required behavior in accordance with his "principle of non-identity," the denial of any equating or necessary linking of word and thing. Piaget showed that only a rare child might grasp this principle by the age of ten or eleven, and then only incompletely.

Piaget said children's beliefs resist attempts to change them before their time. Korzybski said young children could and must learn sound principles of evaluation.

> *Any* identification, at *any* level . . . becomes the foundation for *general* improper evaluation, and, therefore, *general* lack of adjustment, no matter whether the maladjustment is subtle as in daily life, or whether it is aggravated as in the cases of schizophrenia.

Piaget saw children's beliefs as natural and gradual stages of cognitive maturation. Korzybski regarded the beliefs as a reversal of the natural order of evaluation and attributed them, or at least their persistence in adults, to linguistic structures handed down to us from ages when animist, "realist," and "nominal realist" beliefs were the adult standard, just as they are in some tribal cultures today.

I draw from both Piaget and Korzybski. I look to Piaget for insights into children's thinking and to Korzybski for insights into adults' thinking. I rely less on what each said about areas in which he didn't specialize.

You might expect Korzybski's 1933 study, *Science and Sanity*, to reflect Piaget's discoveries. Its bibliography lists four books by Piaget, including ones from which I've quoted earlier. However, I find no evidence in *Science and Sanity* or elsewhere that he used them or even, for that matter, read them.

This may not be as odd as it sounds. Although he accused adults of attitudes toward words and thought resembling those Piaget found in children, Korzybski used different terminology and started from an entirely different point of view. He addressed adult behavior in terms of a grand theory of his own design about levels of abstraction, how linguistic structure leads us astray, and what we can do to get more control over our evaluations.

Korzybski's emphasis (along with that of linguists Whorf and Sapir) on language as a *molder* of thought rather than just its conveyor has entered the mainstream. Few today would argue that it makes no difference how we talk about things. Almost everybody recognizes language as instrumental in behavior. Yet even those most involved in trying to change language habits (for example, people trying to eliminate sexist language in textbooks) may never have heard of Korzybski, Whorf, or Sapir. That's what being mainstream means.

Ironically, one aspect of Korzybski's work that has created continuing difficulty is what he called it. Korzybski adopted the name "*general semantics*" almost as an afterthought, and it's proven a questionable choice. He wanted to encompass physiological, neurological, and psychological aspects of meaning beyond the boundary of narrowly linguistic "semantics." He expected the difference between semantics and general semantics to be observed.

However, no one person or group—and certainly not semanticists—controls where English goes. Contrary to Korzybski's wishes, one meaning of "semantics" has come to be "general semantics." You can look it up. "Semantics" now has both its old, narrowly linguistic, meaning and the new, broadly cognitive-neurological meaning of general semantics. Unlike some Korzybskian purists, I believe the rabbits can't be put back into the hutch; we must live with both meanings. When President Bush replied to a reporter's question by saying that whether foreign citizens trapped in Iraq and Kuwait were hostages was just a question of semantics, I didn't flinch. I presumed that all he meant was let's see what happens before we pick a label.

As Piaget's work on child cognition clearly goes beyond the meanings of words, so does Korzybski's general semantics. Indeed, Korzybski staked out a vast area (some would say too vast) covering human evaluations at all levels. His three-level sequence (event, object, symbol) puts even the lowest verbal level, a simple name or label, at a double remove from the event level. More abstract verbal levels (for example, as "fruit" is to "apples and oranges") will be still further removed. Yet, curiously, the most abstract verbal levels of all, those involving terms like "matter," "thing," "time," "cause," "effect," "choice," "reality," and "meaning," come around full circle to give the adult view of the lowest, or event, level. Therefore, our view of everything, for better or worse, goes through human neural and linguistic structuring influenced both by our own specific experiences and our particular language.

"We do not realize," said Korzybski, "what tremendous power the structure of an habitual language has. It is not an exaggeration to say that . . . the structure which a language exhibits . . . is automatically projected upon the world around us." A formal example would be early scientists classifying "heat" as a material substance (phlogiston) appropriate to its nounhood and then vainly searching for that substance. Commonplace examples include blaming scapegoats for society's ills because "*somebody* must have caused them."

The linguistic structure most generally distorting our view of the event level within Indo-European languages, according to Korzybski, is their subject-predicate structure. This structure dictates that actions must be taken by subjects, as in "It may rain." We accept this structural "it" without qualm, even though on closer scrutiny we can't point to anything at the object (constructed image) level or imagine anything at the event level that we're talking about.

But no single feature causes more harm, according to Korzybski, than our language's "is" of identity. "What 'is' that?" "It 'is' a ball." To Korzybski, this "is" of identity introduces and continually reinforces "identification," the treating of verbal, object, and event levels as equivalent. No one escapes this conditioning.

That the *structure* of our modern languages accommodates animism can't be denied. After "it" starts raining you might hear a frustrated friend say, "My umbrella doesn't want to open." "Don't fret," you reply reassuringly, "the sun's trying to come out." Both sentences exhibit the animist view that umbrellas and the sun are alive and have intentions, don't they?

Of course they do. Let's not deny it.

Few of us note the animism until it's pointed out. Then we try to pass it off as of no consequence, arguing that it's just a manner of speech, that it doesn't mean anything.

Have you ever tried confronting prejudiced people about their language habits? If so, perhaps you sensed how the *way* they talked per-

pctuated their bias. You sensed it, but they didn't. To them their particular words were just a manner of speaking that didn't mean anything.
They may even have told you to stop making a fuss over *nothing*,
thereby clearly signaling that they didn't appreciate your point of view.

Now, generalize from that experience, and you can sense where
Korzybski stood. Lonesome ground. He could hear how almost everything people said was warping their outlooks. The marvel is that he
kept on trying to tell them. That's what we call "being a glutton for
punishment."

Of course, Korzybski found what you'd expect: Most people don't
want their evaluational errors pointed out. They resent humorless criticism of their talking and thinking habits as psychopathological. They
resent being told that children could do better. Yet that's what
Korzybski said.

> Few of us realize the unbelievable traps, some of them of a
> psychopathological character, which the structure of our lan-
> guage sets before us.
>
> The new orientations are simpler than the old because
> they are closer to empirical facts, and so are more easily ab-
> sorbed by children.

The easiest path for most, of course, has been to ignore Korzybski's
general semantics. My wife's psychology text mentions neither him
nor his field. In that book, Korzybski is a nonperson; general semantics, a nonfield.

However, an influential minority, including S. I. (later Senator)
Hayakawa, Irving Lee, Stuart Chase, Wendell Johnson, Elwood Mur-
ray, Anatol Rapoport, and others spread the word widely from the later
1930s into the 1950s. Their books sold in the millions.

By 1972, however, Hayakawa was saying that general semantics had
failed to achieve its idealistic goal of communicating semantic ideas to
the entire community, had failed to develop its theory for the age of
television, and was no longer even fashionable. At the same time he

said that he and other teachers were still getting letters from students stating that the general semantics course they had taken two or three decades earlier was the turning point in their lives.

From these considerations, Hayakawa concluded that general semantics in the teaching profession was particularly suitable for the transition from adolescence to adulthood, for the late high school into early college years, and that thirty years later some students would realize what they had gotten from the training.

You never can tell who might have been influenced by general semantics. For example, I was surprised on finishing the talk on which this book's first chapter is based, given at the 1988 general semantics conference at Yale, to be followed at the podium by an unscheduled walk-on speaker, Alvin Toffler, best-selling author of *Future Shock* and *The Third Wave*. He proceeded to list the Korzybskian points that had most affected his outlook. None of us regulars had known of his interest.

Four years after Hayakawa's 1972 review, maverick educator and author Neil Postman, of *Teaching as a Subversive Activity* (with Charles Weingartner) and *The Disappearance of Childhood*, took on the editorship of the general semantics journal Hayakawa had founded. He later wrote in *Conscientious Objections* that general semantics has been dismissed as *merely* therapeutic by scholars who might have benefited from the therapy and that Korzybski's ideas "are usually labeled 'obvious' by those whose behavior shows the least evidence of their being understood."

If you're looking for intellectual game, Korzybski offers an easy target. Postman said in the laudatory chapter just quoted above that Korzybski's "reach exceeded his grasp," that "*Science and Sanity* is filled with unsupportable assertions and not a few errors, some of them extraordinarily naive." (It's also filled with annoyingly novel punctuation and abbreviations.) Another defender, Anatol Rapoport, of "prisoner's dilemma" and other fame, asked about Korzybski, "If he was not a crackpot ... why was he so repetitive, verbose, pugnacious, redundant and self-congratulatory, manifesting all the symptoms of crackpot delusions?" A campus joke from the 1940s went, "Do you know the difference between *ceramics* and *semantics*? Semantics is *crack*pottery."

Yet Postman, Rapoport, and other general semanticists accept Korzybski's argument that specific training should be provided to help each of us overcome the effects of ancient presumptions embedded in language. Otherwise we are prey to childhood misconceptions, the structures of language supporting those misconceptions, and anyone finding those misconceptions useful in controlling us.

We can't bypass childhood. We can, however, revisit its cognitive beliefs and recognize the folly of thinking we've completely outgrown them.

We can't outlaw those beliefs. Textron, for example, announced a new subsidiary, Textron Lycoming, with full page ads stating boldly, "The beginning of wisdom is calling a thing by its right name," then in very small type, "(Old Chinese proverb)." Yes, very old. Also, tenacious.

Nor can we greatly alter our language's structure. Korzybski prescribed semantic patches (indexing, dating, the etc., quotes, hyphens). I've offered the "*and then?*" These devices, employed as silent habits, can greatly facilitate liberation from childhood's beliefs. But just changing one's language won't insure sound semantics.

For convenience I'll use Korzybski's term "intensional [with an "s"] orientation" to refer to the semantic beliefs Piaget found in children. These include the belief that one can know the essence of a thing from what it's called, the confusion of *name* with *thought* with *thing*, and the belief that the world conforms to our words about it.

"Intensional" sounds exactly like "intentional," but means something different. If this ever trips you up, credit the tenacious hold words have on our thinking. I remember the difference by letting the "s" stand for *subjective semantics*, the kind of orientation that puts undue trust in symbols and signs, that says something must exist because it has a name.

General semantics training seeks to replace an intensional orientation with an extensional one, that is, one that remains aware of the non-identity of events, objects (constructed images), and symbols, and that resolves difficulties by checking the world beyond words.

General semantics training promotes a kind of word-object-event mastery. I still remember feeling liberated at age sixteen to learn from Hayakawa's *Language in Action* that dictionaries are compiled from what authors have written. So that's where the meanings came from! I presume I was already well beyond the belief, now blocked by adult's amnesia, that names inhere in things or come from God. But Hayakawa's words did erase the last traces of any conscious transitional belief I may have had that names somehow related to the essence of things.

Unfortunately, just reading Hayakawa didn't save me from confusing my intensional expectations with the extensional world. Cutting down on blunders took practice.

Even being, as I was at age twenty-three, executive secretary of the International Society for General Semantics didn't save me. To take just one painful example, I once embarrassed an Illinois University professor by sending all his campus colleagues, at his request, membership invitations bearing his name as sponsor. Each invitation arrived at the campus post office boldly stamped INSUFFICIENT POSTAGE. The postal clerk took it upon himself to ask the professor's colleagues one by one as they came by whether they wished to pay the balance due or refuse delivery, explaining as best he could what was inside the envelope. The professor was mortified. His position would have taxed Noël Coward.

This Dickensian scene didn't arise because I'm more than ordinarily irresponsible or foolish. It came from my slipping into an intensional orientation. I'd taken the invitations to the main Chicago post office, asked a clerk to weigh a few to determine the postage needed, bought the stamps, gone off to affix them, and then put the whole lot through a mail chute. My mistake was presuming that the preliminary weighing at *10* A.M. by clerk$_1$ on scale$_1$ *without* stamps was equal to a final weighing at *10:30* A.M. by clerk$_2$ on scale$_2$ *with* stamps. Had I dated the weighings, indexed the clerks and the scales, and asked "and then?" before putting the lot into the chute, I might have gone back to the clerk who first weighed them.

We saw in the previous chapter how intensional reasoning, based on Piaget's findings about children's cognitive habits, could address apples-oranges mathsemantic problems in terms of essences. Let's now repeat the exercise from an extensional point of view, based on Korzybski's general semantics.

Korzybski's extensional approach gives primary importance to the event level (whatever's out there), which must then be distinguished from the object level (what we "see" inside our heads) and from the verbal level (how we talk about it).

The apples-oranges addition problem then becomes, "How do I express at the verbal level what I get by putting two apples and five oranges together at the event level?" We obviously can accomplish the combining at the event level. That's easy. Just put all the apples and oranges in a sack. This leaves the problem of what to call what we've now got in the sack. There's no one necessarily correct answer for all circumstances, yet only a few answers work, and "seven fruit" certainly qualifies.

Similarly, we approach the trips problem extensionally by asking how we could combine one-way trips with round trips at the event level. The answer is by taking two one-way trips to complete a round trip. "But," someone objects, "each one-way trip may not be half a round trip; the traveler may never return."

"True," we reply, "but then no two round trips are the same, no two apples are the same, no two anything are the same. If you follow that line of reasoning, you couldn't add anything."

Given a free voice in naming the units to be counted, we'd probably opt for breaking each round trip in two and reporting one-way trips. That's the airline industry's usual convention. But where round trips are called for, we cut the one-way count in half.

Regarding "passengers," Korzybski's extensional approach requires that we first ask what we mean by "counting passengers" at the event level. We discover agents counting tickets or boardings. We can add all of a given kind of such "passengers" together. Our sum does not represent people, however, but only certain repetitive events.

The same kind of reasoning will lead us to avoid the sorts of nonsense reported earlier that arise from treating counts of "callers," "customers," "patients," and "students" as counts of people. We must start with how we're counting and then find a label that doesn't misrepresent those counts. If we're counting telephone calls, store transactions, patient visits, and one-year school enrollments, we need to keep that in mind and perhaps use those names. Objectifying these repetitive events as people leads to trouble.

An intensional orientation either equates word, object, and event, or projects from one to the next in that order. Thus, knowing something's name takes priority and supposedly promises to tell all or most of what one needs to know about it.

Got that? Now reverse the flow.

An extensional orientation asserts the non-identity of event, object, and word, and constructs knowledge by proceeding in that order. Thus, knowing how something works takes priority and provides a basis for picturing it and giving it a safe name, even if you have to label it yourself.

Christopher Columbus erroneously thought he had sailed to the Far East, the land of Cathay, fabled Cipangu, and the Indies. Columbus remained, as his biographer Samuel Eliot Morison wrote, "stubbornly and obstinately, to the end of his life, absolutely and completely wrong." That's why we have a West Indies and an East Indies, two kinds of "Indians," and attendant unnecessary confusion.

> **Proposition 6:** To improve a map, you have to adjust it to agree with the territory.

It seems that sound mathsemantics requires an extensional, get-beyond-the-words, approach. Piaget's studies indicate that children start with an intensional, words-tell-all, approach. Korzybski's work indicates that many, if not most, adults fail to make a solid shift from the childhood intensional to a fully adult extensional orientation.

These thoughts lead to the conclusion that most adults don't automatically develop sound mathsemantics.

Hayakawa's work indicates that an educational correction of the failure to shift from an intensional to an extensional orientation works best at the end of high school or the start of college. Such a correction seems rather late for mathsemantic development. I think we must start earlier.

I find Piaget too accepting of childhood semantics. I agree with Korzybski that we can and should guide children away from the pre-scientific semantic presumptions embedded in our language. We can provide such guidance with utter gentleness and simplicity. Witness a parent's report from the October 1991 Institute of General Semantics *Newsletter*.

Rachel began her third year of life like most two-year-olds, asking the time-honored questions: *"What's dis?"* and *"What's dat?"* Faced with my first crisis, I bravely ventured, *"We call that a truck."* After a few months of these gentle answers, I was rewarded with a new question, *"What call it, daddy?"* Progress! Not wanting to keep all the fun to myself, I began to ask her what *she* calls it. . . .

But all was not smooth-sailing. At thirty-six to forty months, Rachel starting playing more with language. After hearing my answer, *"We call it red,"* she would exclaim with a grin, *"I call it blue!"* Now we've got trouble. I agreed with her that she *could* call the truck blue, but suggested her friends might not hand it to her, as most others would call this particular truck red. This engendered no response. Perhaps independence held a greater fascination for her than conformity.

During the last six months an interest in functionality has emerged. While the majority of her questions still retain the usual subject-predicate structure (e.g., "What is this?"), she has also begun asking *"What it do?"* and *"What do with it?"* I'd like to believe our earlier interactions led to this development, but I'm not able to make any independent determination.

Richard Feynman writes in *"What Do You Care What Other People Think?": Further Adventures of a Curious Character,*

> On weekends my father would take me for walks in the woods and he'd tell me about interesting things. . . .
>
> The next Monday . . . we kids were playing in a field. One kid says to me, "See that bird? What kind of bird is that?"
>
> I said, "I haven't the slightest idea. . . ."
>
> He says, "It's a brown-throated thrush. Your father doesn't teach you anything!"
>
> But it was the opposite. He had already taught me: "See that bird?" he says. "It's a Spencer's warbler." (I knew he didn't know the real name.) "Well, in Italian, it's a *Chutto Lapittida*. In Portuguese, it's a *Bom da Peida*. In Chinese, it's a *Chung-long-tah*, and in Japanese, it's a *Katano Tekeda*. You can know the name of that bird in all the languages of the world, but when you're finished, you'll know absolutely nothing whatever about the bird. You'll only know about humans in different places, and what they call the bird. So let's look at the bird and see what it's *doing*—that's what counts." (I learned very early the difference between knowing the name of something and knowing something.)

In addition to this semantic insight, Feynman's father also introduced him to a world of order, structure, and process.

> Another time, when I was older, he picked a leaf off of a tree. This leaf had a flaw . . . a little brown line in the shape of a C, starting somewhere in the middle of the leaf and going out in a curl to the edge.

"Look at this brown line," he says. "It's narrow at the beginning and it's wider as it goes to the edge. What this is, is a fly—a blue fly with yellow eyes and green wings has come and laid an egg on this leaf. Then, when the egg hatches into a maggot (a caterpillar-like thing), it spends its whole life eating this leaf—that's where it gets its food. As it eats along, it leaves behind this brown trail of eaten leaf. As the maggot grows, the trail grows wider until he's grown to full size at the end of the leaf, where he turns into a fly—a blue fly with yellow eyes and green wings—who flies away and lays an egg on another leaf."

"That story," Feynman says of a similar lesson, "was probably incorrect in *detail*, but what he was telling me was right in *principle*."

We don't know how Rachel will turn out. Jeff Mordkowitz, Rachel's father and a Trustee of the Institute of General Semantics, wrote his report in 1991, before Rachel was four.

Richard Feynman (1918–1988), whose father probably never heard of general semantics, once with outrageous candor—to the dismay of the U.S. State Department—told a lecture hall full of Brazilian physics students, physics professors, and government officials, that "the main purpose of my talk is to demonstrate that *no* science is being taught in Brazil." He then told how struck he had been on arriving in Brazil to see all the science classes they had and how even elementary school kids bought physics books. But, he explained, the books and instruction hadn't worked. "Have you got science? No! You have only told what a word meant in terms of other words." Feynman, of course, won the 1965 Nobel Prize in Physics.

We can now build on proposition 6.

Proposition 7: To improve our mathsemantic maps, we must learn to think extensionally.

CHAPTER **6**

More problems with names

N ames complicate addition. They devastate multiplication and division.

Mathematicians don't talk much about names. Their term is "units." To confuse matters, they refer to several different things as units. For example, the number "one" is a unit. That makes etymological sense; for "unit" is a back-formation from "unity," or so my dictionary tells me. But then any number can be a unit; we can count off in units of two, or three, or whatever.

Numbers and units wobble our grammar. In counting guests for tables of bridge, for example, four make (or should it be "makes"?) a unit. Going on, the number to the left of the decimal point (e.g., the "3" in 3.2) is the unit (or should it be "units"?) and that left-side column is the units (or should it be "unit's," let us hope not "units's"?) column.

Then, when you're counting, whatever "one" "is" of whatever you're counting (e.g., one egg) is a unit. That's the sense in which "unit" meets "name." Any standardized measurement can also be used as a unit (mile, life, tablespoon, gram, etc.).

"Unit" has an aura of mathematical precision about it. It sounds like something precise and beyond question. Except it isn't. That's why I talk about "names." It keeps me reminded that we fallible people are involved. Nevertheless, sometimes I'll use "unit." It's a good name for "one of something," if you forget the math aura and remember that we named the something.

Now, our one hundred ninety-six clerical applicants proved reasonably "good-at-numbers" in many kinds of plain number multiplication. Nineteen out of twenty (188) applicants correctly did "171 times 3." Four of five (159) correctly did "1246 times 26." Three of five (120) correctly did "1.407 times .32," decimal point and all. Just over half (99) correctly did "−3 times 4.2," handling both the minus sign and the decimal point. They answered "513," "32,396," "0.45024," and "−12.6."

But only eleven applicants correctly handled a multiplication problem involving names.

16 travelers
2.5 hours
_____

You'd certainly expect at least the one hundred twenty applicants who'd correctly multiplied two numbers in decimals to handle 16 times 2.5 with ease. The trouble, then, must have come with the travelers and hours.

As evidence of the confusion this problem created, the one hundred ninety-six applicants gave sixty-six digitally different answers to this problem. Once again, for your convenience, I'll group their answers into manageable sets.

Eleven applicants, as already noted, or just one in eighteen, gave the answer desired, 40 traveler-hours. The "16" times "2.5" yields "40." The "travelers" times "hours" yields "traveler-hours."

Note how this correct answer tracks the common business calculation of "men" times "hours" to get "man-hours." My desk dictionary defines "man-hour" as "a unit of one hour's work by one man used esp. as a basis for cost accounting and wages."

Nine other applicants came close. They gave answers that approached the right idea, for example, "40 hours traveled," "40 traveling hours," and "40 hrs traveling time." Such answers seem to show a good brain working on a new problem and just not quite solving it. There's a difference between a traveler-hour and an hour traveled, if only that a traveler-hour is timeless while an hour traveled refers to a completed event. Perhaps you can hear the difference better in comparing man-hours with hours worked (hours manned).

The convention, a good one, followed in English for multiplying differently named units is to hyphenate them. As in compound nouns generally, the first part stays singular; the second part can take a plural as appropriate. One man-hour, two man-hours. In estimating heating requirements: one degree-day, two degree-days; in water reservoir volumes: one acre-foot, two acre-feet; in museum attendance loads: one visitor-hour, two visitor-hours. Thus, one traveler-hour, two traveler-hours.

I'm not sure whether there's any rule regarding which of the two terms should come first. You can probably let your ear judge. Certainly man-hours sounds more reasonable than hour-men.

Before we get too far away from the problem, here are the other answers given. Seventy-two applicants, the largest single group, amounting to four of eleven, gave just a number, no unit. These answers varied from −1 to 400 and included all these numerically incorrect answers: 4, 30, 32, 32.5, 32.8, 33, 36, 38, 40.6, 112, and 270. Don't ask me where these answers came from. I don't know. Most of these people who gave no units in this answer had already bombed out on other questions, and I didn't get to talk with them. Perhaps they just flipped when they saw they had to multiply travelers by hours.

However, even if they flipped, why did they land on these particular answers? I wish I knew. I believe we should study such errors for the insights they may provide into people's mathsemantic beliefs. Looking away from error in embarrassment makes no sense to me. Think, for example, where medical science would be if we squeamishly shunned injuries and diseases. We'd never get anywhere. I think we need to look carefully at what ails us. Just marking answers right or wrong to get a total grade leaves me dissatisfied; I want to know *why* someone would answer −1 or 112 or 270. What helpful mysteries might those answers reveal if only we knew how to interpret them? Sadly, about these particular answers, I haven't a clue.

Forty-five applicants, or about two in nine, seem to have decided that travelers don't count. They expressed their answers just in hours, from 2.5 hours to 40.4 hours. One possibly revealing answer was, "40 hours if consecutive." This seems to relate somehow to the difficulty of objectifying the answer, of trying to pull it together.

Only three applicants ignored the hours and expressed their answers just in travelers. But "40 travelers" is wrong, "16 travelers" is worse, and "4.00/traveler" is baffling. It seems to say "exactly four per traveler." But four of what?

Forty-three applicants, or almost two in nine, gave no answer. Only two said the problem was impossible, and one of the two spelled it out clearly, "can't multiply two unlike things." This echoes the apples-oranges refrain, "can't add two unlike things." And it's just as wrong. If no two things are the same, then even taking just two things means taking two *different* things. We also can multiply different units, passengers times hours, just as the problem asked.

The remaining eleven applicants, one in eighteen, managed to work both travelers and hours somehow into their wrong answers: "4.0 travelers hours," "32 hours in travel," "37 total hours traveled," "40 travelers in 2 1/2 hours," "40 travelers per hour," "40 hours/traveler," and "46 traveling hours," among others. Note that both "travelers per hour" and "hours per traveler" occur.

Now, scrunch down for a moment and pretend you're one of Piaget's children. Concentrate on believing that the essence of things is given by their names. Then you must sense that multiplying travelers by hours calls for some kind of name. But let's say you don't know what that name is. You're left adrift; you're missing the essence.

So, still thinking intensionally, you try to call up an image on the object level. What happens? You can picture a passenger; you see a person. You can call up an hour, perhaps as hands on a clock. But how do you multiply them? "What," you might ask, "does multiplying them mean?" You don't even know whether the hours are consecutive.

Lacking a name and lacking a coherent image, the intensional approach gives no immediate way to work the multiplication. So you look for a rule, another intensional device. Unfortunately, you can't remember ever having heard of a rule for this case. You begin to wonder whether you missed school the day that was covered. Time now presses, so you stab at an answer and move on.

I don't know how many applicants played out this scenario. But I can remember approaching problems, and not just math problems either, that way. Maybe you can too; maybe not.

In any event, the extensional approach works differently. Recall that the extensional approach solves the apples-and-oranges addition problem by working from the event level to the verbal level. First we imagine adding things physically, try to see them in our mind's eye at the object level, and if we succeed, we then have left only the question of what to call the collection.

The extensional approach works the same way with multiplication. First we have to imagine multiplying travelers by hours at the event level. What does "multiply" mean in this case? Well, even in plain arithmetic, "two times two" means taking two twice. Similarly, multiplying a traveler by an hour means taking one traveler one hour. In idiomatic English, taking one traveler *for* one hour. Picture that event at the object level, one traveler, say, on a plane, for one hour. What would you call that event?

The first time around, you may not know. That's understandable. But, if you start with that picture of an event and either ask or stay alert, you'll soon learn our language's convention for such events. It expresses "one traveler for one hour" as "one traveler-hour." You can use this convention for as many new multiplications as you like: car-miles, passenger-miles, work-years, pet-years, television-viewing-hours. "Our two cars produce 3,000 car-miles per month." "The people in this office have put in a total of 247 work-years with the company."

What's involved here goes by the name "dimension." Most people have come across dimensions on a rudimentary plane in geometry. They've multiplied length times width, for example, when both were expressed in feet. That gave feet times feet, which by mathematical convention is called "feet squared," or "square feet." They may also have run into three dimensions in multiplying length times width times height to get volumes, perhaps expressed in cubic feet, which, as the units for *feet* times *feet* times *feet*, surely beats "foot-foot-feet."

People are generally less familiar with dimension just when it gets interesting and starts to involve other kinds of units. They may have heard of time as "the" fourth dimension. They've probably never thought of a passenger or a car as a dimension.

Yet passengers and cars provide dimensions. My desk dictionary defines dimension in part as follows: "the range over which or degree to which something extends, scope." Thus, if you want to think of a mile as extending to one car, or a car ranging over one mile, you're using both "car" and "mile" as dimensions. You can multiply them. Their unit-product, a "car-mile," is the movement of one car for one mile. A passenger-hour is the duration (and perhaps the endurance) of one passenger for one hour.

A much easier problem, but one that still gives people lots of trouble, is dividing units by the same units. Because our work used such divisions, we included the following question in our quiz.

$$30 \text{ min.} \overline{)\,2 \text{ hrs. } 12 \text{ min.}}$$

The one hundred ninety-six applicants gave a startling total of eighty-two digitally different answers to this division problem. As usual, I'll group them for easier handling.

Fifty-eight applicants, or about three in ten, gave the desired answer, "4.4" (or "4 2/5," and we also allowed "4 6/15," and "4 12/30," although the usual convention is to reduce such fractions). I don't know what method applicants used to get their correct answers. Several methods work.

Applicants may have started by simplifying, by asking themselves

how many times thirty minutes goes into two hours. I guess almost everybody could answer that: four times. They could then ask how many
times thirty minutes goes into twelve minutes. The answer here is
clearly something, but less than once. They would then know that the
complete answer was between four and five.

The exact answer requires dividing 30 into 12, which they could do
longhand or express as 12 divided by 30, which is 12/30, which reduces to 2/5, which can also be expressed as 0.4. So the final answer
is 4.4 or 4 2/5, whichever you prefer.

Alternatively, applicants could convert everything into minutes from
the beginning. The problem then becomes 30 minutes divided into 132
minutes, which can be written as 132 min./30 min., which reduces to
132/30, because dividing time into time (like a half hour into a whole
hour) is expressed as a number. The fraction can then be reduced in
any of several ways. One way would be to divide first by ten to get
13.2/3, and then divide by three to get 4.4/1, which equals 4.4.

They could also get the answer by long division. The division is simpler than others the applicants solved.

$$30 \overline{)132}$$

The important thing, however, is getting the units right, otherwise
we don't know what we're talking about. Because two hours divided by
30 minutes is simply four, the answer has to be a plain number, no
units, between four and five.

The largest single group of answers, seventy-six in all, however, expressed the answer in terms of time. These said, in effect, that a half
hour goes into two hours not four times but four "hours" times, or perhaps four "minutes" times. To me that's nonsensical, I can't visualize
it. Here's a sample of these answers, low to high.

.44 min
2.4 min
4 min 4 sec

4 min 12 sec
4 2/5 min
40 min
40.4 min
44 min
1 hr 2 min
1 hr. 48 min
120 min
132 min
2.4 hrs
3 hrs 22 min.
4 hours
4 hrs 24 min
4.4 hours
8 hours

From an extensional point of view, none of these answers makes sense as an answer to a question dividing time by time.

"Tell me, friend, how many half hours are there in two hours and twelve minutes."

"Well, let me see now. Could it be four minutes and four seconds? That's not it? Okay, well then, how about one hour and two minutes? No? Eight hours?"

I apologize for poking fun at these answers. I'm actually a little embarrassed, for I wish I'd taken the time to investigate the mathsemantic outlook behind them. I'm sure they're mostly brave attempts to solve what must have looked like a deep mathematical puzzle.

But I'm equally sure there's almost no one reading this book who can't figure out there are four (not four minutes or four hours, but just plain four) half hours in two hours. Anyone able to figure that out and not panic would then be able to see that the answer to our division problem has to be a plain number somewhat greater than four.

With an intensional orientation, however, one could look at the original problem, fail to see what kind of a name the answer would have, find it impossible to visualize, fail to come up with a rule, and therefore be reduced to taking a stab and moving on. That's apparently what a lot of our applicants had to do.

Thirty-four applicants gave answers in plain numbers (including fractions) that were wrong. These ranged from 5/22 to 132. While they didn't attach the time dimension to their answers, these applicants were clearly confused by it. We know they were troubled by the time-units aspect, and not just by division, because the overwhelming majority had already solved three earlier division problems not involving units.

Eight applicants gave answers in figures that were unlike any ordinary number I've ever seen. Examples ranged from "4   0" to "706   20/." More confusion. Two applicants equivocated and eighteen gave no answer.

The evidence seems to be mounting that a lot of people who can use words or do math reasonably well can't do both at the same time. Pushed into trying, they get rattled. It's as if they see the mathsemantic no-man's-land ahead, recall being there before, remember mine fields, but don't know how to pass through safely.

If that's the case, it means people more or less recognize the mathsemantic boundary. That helps a lot. It gives a starting place. So, if we get queasy when a mathsemantic problem looms ahead, we should be thankful for the warning.

The rule to follow on getting the warning is this: Solve the semantic problem first. Go extensional. Ask what events "out there" could possibly correspond to the words.

Let me give you an example, actually a rather famous example. You'll recall that I promised to put no useless puzzles in this book, nothing designed to show what marvelous things math can do. I'm going to keep that promise. Nevertheless, here comes a puzzle. I can do this honestly because this puzzle will be useful in that it will show what alone math *can't* do, what you must use semantics for.

First, the puzzle.

Find the Extra Dollar
Three men came into a hotel. A room with three single

beds cost $30.00. Each man gave the hotel clerk $10.00. After the men had gone to their room, the hotel clerk realized he had overcharged them $5.00 for the room. He called the bell-hop and gave him $5.00 to return to the guests. On the way to their room, the bellhop decided he would keep $2.00 and give each man $1.00. This meant that each man paid only $9.00 each. The room cost them only $27.00. The bellhop kept $2.00. Where is the extra dollar?

Perhaps you recall seeing this puzzle before. I got this particular rendition of it from a book on speech principles published in 1982 by a leading textbook publisher. Before I show you the way to solve the puzzle, please read next what the teacher's guide had to say about it. This is, in effect, what the book suggests teachers tell aspiring speech students.

It depends on how you count. Try counting to ten on your fingers, then count backwards from ten to one. When you count backwards from ten to six on your right hand, there are still five fingers on your other hand which when added to the six equals eleven.

I hope this explanation leaves you more confused than before you read it. It did me.

Fortunately, I was taught not to believe everything I see in print. Also, to say I don't know when I don't know. So after reading this explanation, doing what it said, reading it again, and still not seeing how it might help me, I decided to strike out on my own.

If you're coming with me, what that decision means is getting extensional about the puzzle. Let's start by stripping the decimals and zeros. Their function here, I bet, is just math-precision one-upsmanship. We have then a plain picture of three men who have just spent $27 at a hotel. Where did their $27 go? The desk clerk took $25 for the room and the bellhop took $2 for himself. $25 plus $2 makes $27. That's right, isn't it? Hey! Where's the problem?

You'd like to trace the money further? Good. The men gave $30 to the desk clerk. Okay? The desk clerk gave $5 to the bellhop to return to the men. Okay? The bellhop kept $2 and gave $3 to the men. Okay? $2 plus $3 makes $5. Okay? That's straightforward, wouldn't you agree? So, where's the problem?

Indeed, just as we might have nothing to fear but fear itself, we seem to have no problem in this problem except finding the problem.

The problem is semantic. It's in the last three sentences of the statement of the puzzle. The first sentence is: "The room cost them only $27.00." That depends on how you look at it, doesn't it? The desk clerk, for one, thinks they paid $25 for the room. The bellhop wants the men to think the room cost $27, but that's a deception. The men may think that, but the bellhop knows better.

A truer statement, therefore, might be, "The room cost them only $25, and the bellhop stole $2, so the men were out $27." Put this way, the puzzle's next sentence, "The bellhop kept $2.00," is merely a redundancy. The implication that the $27 and the $2 should be added is false. We've already counted that $2 in the $27. Adding it again makes no sense.

Neither does the final question, "Where is the extra dollar?" The question implies that the $29 reached through redundant addition should really total $30. Well, why should it? The $29 is a figure that makes no sense. There's no extra dollar.

The math side is simple. The semantics side is where the trouble lies. The problem starts with the phrase, "The room cost them . . ." We too quickly objectify *room*. Then we think of the bellhop's theft as separate, perhaps as a kind of involuntary tip. But the theft has been included as part of the so-called room cost. That's the puzzle's game, to trip us up on semantics while intimidating us with some presumed sanctity of math.

What saddens me most about this puzzle is the guidebook's explanation. Rather than helping speech teachers and students, the explanation seems designed to mystify the math-speech boundary. The "explanation" doesn't *explain* anything. It just adds to the confusion, which raises a pedagogical question. Why wouldn't a speech textbook solve the semantic problem rather than confusing it with a mathematical one?

Proposition 8: In mathsemantic problems, it pays to tackle the semantics first.

CHAPTER

Three domains

You may have heard it said that mathematics is a language. I can remember being told that more than once, starting perhaps in fourth or fifth grade. I now realize how little I understood what it meant.

I thought I knew what a language was. My mother did her sums in French; I knew that French and English were different languages. But like Piaget's children, I didn't make much of the distinction between words and things. Therefore, I saw neither the importance of math's being a language nor just how odd a language it is.

Korzybski, you'll recall, considered the event, object (mental-construct), and verbal (symbolic) levels of abstraction as three quite separate domains. He deplored our mistaking one for another. But besides the obvious fact that the word is not the thing and the map is not the territory, what sets language apart from the event and object levels?

Let's look first at how the levels connect.

As the event level rolls along, doing whatever it does, our sense organs can follow it like TV cameras producing live, "real-time" images. The two-dimensional images on the screen obviously aren't the real events. Yet changes in events produce related changes on the screen. The TV, in computer parlance, produces real-time analogs of events. Our sense organs also produce real-time analogs, at what Korzybski called the object level.

Language, however, relates to events and their object-level analogs predominantly in an off-line digital mode. "Off-line" means not tied immediately to the event, like a broadcast "recorded earlier." "Digital" means put into code not varying continuously with the event.

Take color, for instance. The event level exudes electromagnetic radiation of various wavelengths. When waves in the narrow band of visible light impinge on our eyes we see color. As wavelengths change, we see analogous changes in hue. But to verbalize what we see, we employ names: blue, green, yellow, orange, red, purple, and so on. The best our words can do to represent continuously evolving events, say a sunset, is poor indeed. Our words can't model the event or anything remotely like it. In practice we don't try. Instead, we use descriptions that rely heavily on evoking memories of past sunsets or something similar.

Thus, when the event-level sunset does what it does, our senses can provide an analog image in real time, but our ordinary language can produce only an off-line digital record.

"But," you may object, "how about a road map? I presume you consider that to be off-line and at the symbolic level. Isn't a map more analog than digital?"

Good point, and that's why I said language is *predominantly* digital. Actually, math helps mightily to reduce language's digital shortcomings. A scientist, for example, can describe colors in numerical wavelengths to give a smoother picture of continuous changes, although perhaps only to other scientists. A road map, of course, depends on geometry.

But even a road map, whose locations do vary continuously as analog elements, will have digital symbols (pictographs, arrows, colors, shading, words, different type faces) to indicate north, built-up areas, bridges, road widths, road access, elevations, and picnic areas. What, except our conventions, makes black and red lines mean roads?

Without the names, symbols, and legends, road maps would be like unidentified aerial photographs, not very useful for getting around.

In general, we need digital frameworks and digital details to make sense of verbal-level representations, even ones like maps, which have analog elements.

Most verbal-level output is almost completely digital. This book, for example. Its little words are not analogs of the big territory they represent. I say "big," but the word is not big. Even the word "little" is bigger.

Being digital is not necessarily a drawback. Chinese ideograms have descended from stylized pictures; many are analogs of object-level constructs. As such, they may help readers remember meanings. However, such ideograms provide no clue to pronunciation. The digital forms of English words do not resemble their meanings, but they do permit alphabetic filing schemes and simplify typewriter design.

Nor is being off-line necessarily a drawback. Imagine a world with only live TV transmission, no VCRs, no movies, no stills, no transcripts, no books, no advance information, and no postmortems. Everything would happen live and just once. There'd be no way to replay anything off-line to get a better handle on it.

So, the verbal level has useful properties not found at the event level and found only in part at the object level. Language's digital coding and off-line performance permit revisitation and communication. Is there more? Yes.

Four verbal-level properties of particular interest to mathsemantics are contradiction, equivalence, classification, and self-reference.

I remember from my New York days planning a drive from our Manhattan apartment to northern Vermont. Because the drive was long, the territory new, and our children young, I took special pains to map the route in advance. Indeed, to be on the safe side, I checked three maps.

An odd by-product of my caution was finding an intersection represented differently on each map. Think of a normal intersection of an east-west and a north-south highway. That was map one. Now keep the roads leading north, south, and west from the intersection as they are, but move the road going east up a little north of the intersection. That was map two. Repeat the process but this time move the road going east down a little south of the intersection. That was map three.

I wondered all the way to the intersection which map had it right. It turned out all three were wrong. The actual road from the east divided just before reaching the intersection and thereby managed to enter both above and below its western leg. Given already cramped maps, three cartographers had chosen to show this in three different ways.

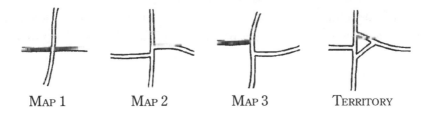

Map 1   Map 2   Map 3   Territory

Now, *even without knowing the territory*, you can see that the first three maps contradict each other. Contradiction is a verbal-level phenomenon. The territory doesn't enter into it. The event level can't contradict itself. Indeed, strictly and etymologically (contradiction means to *speak against*) territory can't contradict maps either. It may, of course, happen not to agree with them.

Let's say we had two atlases. If one offered contradictory maps, we'd tend to avoid it in favor of one whose maps agreed with each other. So, *without knowing the territory*, we can use the absence of contradiction as a basis for choosing an atlas. It may not always be the best choice, but other things being equal (which they never actually are) we'd generally prefer a consistent to an inconsistent set of maps.

Contradiction, then, gives us a basis entirely at the verbal level to choose between alternative formulations.

When our concern is with the territory, we want maps that agree with their territories. In the case above, we would want all three maps adjusted.

When our concern is with language, however, we want maps with internal consistency. And what makes math such an odd language is that it pursues internal consistency almost to the exclusion of any other concern.

The rules of math (its grammar, if you'll allow) and the statements governed by them depend no more on outside events than does the gender of French or German nouns. In math we can set up any system of relationships we want. So long as we observe our own rules and avoid internal inconsistency, we'll be in the clear.

That's what pure mathematicians do. They set up systems within which the correctness of any statement (called "validity" rather than "truth") depends on whether it can be deduced without contradiction from other parts of the system. In this sense, math is only about math. It's a language with no event-level referents. Or, rather, it's an interrelated set of specialized dialects (arithmetic, plane geometry, trigonometry, calculus, group theory, symbolic logic, etc.) all having this characteristic of noncontradiction.

We can and do, however, also consider math as descriptive of events. Our description may then not agree with the events. It can be wrong. When it is, we'll first look to see whether we've violated the math rules. If we've followed the rules correctly and our description still doesn't fit the events, we'll nevertheless not fault math itself. Rather, we'll say that we've failed to classify the events properly or that we've picked the wrong kind of math.

Say we find that one rabbit plus one rabbit produces eight rabbits rather than two, and that one pint (of stones) plus one pint (of water) produces one and a half pints of whatever. Neither these nor any number of additional contrary examples will persuade us to abandon our conviction that one plus one equals two. That's because we regard math as an internally consistent language not to be judged by events.

It follows that one can learn math as a language and still know next to nothing about using it. That requires knowing something about rabbits, water, and stones.

Next, consider equivalence. At the verbal level we can define one word with other words, treating them as equivalent. We can say "the woman's spouse" and mean the same thing as "her husband." To state equivalences with brevity, the language to pick is math. "1 + 1 = 2," "2 × 2 = 4," "$3^3 = 27$."

Such equivalences do not occur at the event level. Every event is itself; no two are the same. If they appear the same to us, that's our judgment of the events, not the events themselves.

Nor do equivalences occur between levels. The map is not the apparent object is not the event.

Next, consider classification. Here's another verbal-level activity. No classification is forced by events. Events just are. We can classify them as we see fit. We can group an orange and a lemon as citrus, or fruit, or tart-tasting, or orange-yellow, or pulpy, or tree-borne, or subtropical, or round, or thick-skinned, or in other apparently natural ways. We can also classify them as liked-by-us, exhibits in a murder trial, mementos, sources of flavoring, and so on.

Classifications of events often cross verbal-object-event boundaries. Take "sound waves." The term labels a physical event (waves) in terms of our object-level perception of it (sound). Question: Does a tuning fork that creates air waves above our hearing threshold emit sound waves? Answer: It depends on how you want to classify them. There's no necessary linkage between word and event. We adults can call the sun the moon, and the moon the sun, if we want.

Any lawyer worth a fee can argue convincingly (1) that any two things are different, or (2) that the two things are really the same, depending entirely on which helps the lawyer's client.

Next, consider self-reference, also known as self-reflexiveness, recursiveness, and recursion. Each term means that language can refer to itself. For example, "This sentence is short."

Now, every adult should be let in on the secret that self-reference is the wellspring of paradox. "This sentence makes a false statement." What strange fun! If the sentence makes a false statement, then the sentence is true. And if the sentence makes no false statement, then it is false.

This recursive property does not, thankfully, occur at the event level. It arises from playing around at the verbal level.

If, like children, we fail to distinguish event level from verbal level, then we face some adult-size headaches. We think the contradictions and paradoxes are "out there." In our confusion, we more easily succumb to demagogues' mistaken equivalences and classifications.

Fortunately, the closest most people get to these headaches is in trying to follow the zany events in *Alice in Wonderland* and *Through the Looking Glass* that mimic language's features.

> "I can't explain *myself*, I'm afraid, sir," said Alice, "because I'm not myself, you see."
> "I don't see," said the Caterpillar.

One nice thing about Alice, she's a girl. She goes through all these mad adventures, finally rebels, and emerges none the worse for them.

For years it was thought that male brains were wired better for math than female brains. Then it began to appear that the male's advantage related not to math in general but primarily to spatial visualization. Recently even this presumed advantage seems to have faded. Being a feminist myself, I long ago ventured the opinion that nurture, not nature, was at work here. "Boys," I said, "have more incentive today to visualize and measure space than girls." The incentives when I was a boy were even more lopsided.

My wife is bright and effective. Yet just the other day while watching the Philadelphia Eagles drop a game to the Washington Redskins she asked me how long a football field was. She'd often noticed the stripes across the field, but hadn't known each marked off five yards. She'd never noticed the one-yard hash marks at all. Perhaps I shouldn't have informed her. She no longer attributes my scooping CBS in announcing play yardages to my naturally superior estimating ability.

Our views of the world contain many questions that remain to be settled. The equivalence of male and female brains or lack thereof is a good example. I expect one day we'll find answers far more interesting than any question we now ask on this subject. Meanwhile, the brain's ten billion or so nerve cells and their enormously greater number of interconnections pose a considerable event-level barrier, a totally connected vermicelli tangle, gloriously alive. No wonder our object-level and verbal-level abstractions have a way to go.

So, we want to keep the levels of abstraction clear. This is harder than it seems. Just when we feel we can't be misled by the "is" of identity, somebody throws a level-straddling term at us and we don't recognize it. Take "history." Sounds innocent. But does it mean the events themselves or the words about them? We need to decide each time. Take "fact." Does it refer to an event-level occurrence, our object-level construction of it, or our statement about it? We need to decide each time.

Language sets plenty of traps like these. It promotes the identification of event, object, and symbol. To slip back into childlike outlooks, to start believing that our words state the essence of things, we need only drop our guard.

I greeted the 1989 publication of *Everybody Counts: A Report to the Nation on the Future of Mathematics Education* with enthusiasm. "Here," thought I, "comes an authoritative book." I don't recall one boasting a longer list of distinguished collaborators and supporters, including the National Research Council, the National Association of Science, the Mathematical Sciences Education Board, the Board on Mathematical Sciences, the Committee on the Mathematical Sciences in the Year 2000, etc., etc.

What I expected was a book about math education produced by math proponents with verbal-level awareness. I expected consistency and language carefully selected to fit events.

"Mathematics," the book began, "is the key to opportunity."

Now it just so happens that my faculty advisor at Chicago in 1947 suggested I take honors work in math, but I declined in favor of broadening my studies and haven't yet regretted it. So, while I accept that learning mathematics may help some people and our society get ahead, I don't agree that mathematics, whatever it "is," is *"the"* key to opportunity. Strike one. I was awake.

"No longer just the language of science," read the second sentence, "mathematics now contributes in direct and fundamental ways to business, finance, health, and defense."

This implied that mathematics used to be just the language of science, which I doubted. So I looked it up. What I found from mathematical sources was that five thousand years ago, the Babylonians used math for trade accounts. Four thousand years ago the Egyptians put mathematics to practical uses such as building pyramids, orienting temples on true north-south lines, and reestablishing property lines after floods. During an especially good year for olives twenty-six hundred years ago, the Greek mathematician Thales is said to have cornered the market by acquiring all the olive presses. (One of his pupils was Pythagoras.) Twenty-three hundred years ago Archimedes used math to design war machines and to determine whether the king's crown had honest proportions of gold and silver. Strike two.

On page seven I read, "It is vitally important for society that *all* citizens benefit equally from high-quality mathematical education." What on earth was meant by "equally"? How could anyone who cared about math use "equal" so sloppily? Isn't equality in human affairs, no matter how desirable, really impossible? And where would it leave us if a vital goal was impossible? Dead, I presumed. I wanted to call strike three, but let it pass as a foul ball.

But then page ten baldly stated "the fact that many adults who never learned mathematics have succeeded without it." What struck me about this conclusion was not its truth, which is undeniable, but its flat contradiction of the page-one opening sentence. Here on page ten were people who had opened the door to opportunity without the *page-one key*. That did it. *Strike three. You're out.*

I read the rest, contradictions and all, sadly, as you might watch a friend's son losing a twenty-to-nothing game.

Proposition 9: Math enthusiasts need to watch their language.

CHAPTER

Grasping nothing

We come now to an unexpectedly deep problem in the recruitment quiz. Applicants found it even tougher than adding apples and oranges.

The problem was one of a group for which the direction was: "Round the following numbers to the nearest whole number." The specific number was .098 (point 098). If you translate such decimals into dollar terms for quick comprehension, .098 amounts to just under ten cents. Ten cents to the nearest dollar would be no dollars, so the answer is zero.

Our one hundred ninety-six applicants gave twenty-three digitally different answers to this problem. These ran from .01 to 98,000, but I'll report on just four groups.

The most popular answer was "1." So said fifty-five applicants, or more than one in four. The next most frequent answer was ".1," given by fifty-three applicants, again over one in four. Then came the answer hoped for, "0" (zero), given by forty-eight applicants, or just under one in four. The only other frequent choice was to give no answer. Sixteen applicants did that, or about one in twelve.

Conversation with applicants about this question revealed that they shunned zero. Whatever they understood about rounding foundered on their feeling that you can't round something to nothing because, well, because nothing is nothing and a whole number has to be something, so nothing can't be a whole number.

I tried a practical example. "Suppose we kept accounts to the nearest dollar," I suggested. "We can do that, can't we?" This was quickly agreed. "And let's say somebody owes us ten cents. How do we post that to the nearest dollar?"

"We post it as one dollar," I was told. At first I thought somebody was pulling my leg, but this is the answer I got. I presume math teachers know this. But it surprised me. I knew of the distaste for rounding to zero; that's why it was on our quiz. Yet I still hadn't expected anyone to round ten cents to a dollar.

So I did what you might do. I asked what one dollar ten cents would round to. Some answered, "one dollar," which opened a discussion on how both ten cents and one dollar ten cents could round to a dollar.

Others answered, "two dollars," which opened a discussion on whether one dollar ten cents really was closer to two dollars than to one dollar.

Both discussions revealed more confusion about zero than I could dispel during an employment interview. Zero wasn't a number to these applicants. It was a symbol of nothingness, of the void, of non-existence.

Piaget said of French-Swiss children, "The idea of non-existence always causes difficulty." I seemed to have found the problem in suburban-Philadelphia job applicants.

Actually, I had stumbled into something far more interesting.

The German historian and philosopher Oswald Spengler (1880–1936) wrote in many fields, including math, science, and art. His major work was *The Decline of the West*, an early statement of the idea that "whatever goes around, comes around." Spengler held that human history reveals not continuous progress along one line but, instead, alternative cultures, each of which either already has completed or must still complete its own life cycle of youth, maturity, old age, and death. Western civilization was no exception.

James R. Newman's commentary in my dog-eared copy of *The World of Mathematics* describes Spengler as "a capable mathematician" whose "disturbing and exciting" ideas "cannot be dismissed as hollow." Among them, with Spengler's emphasis:

> *There is not, and cannot be, number as such.* There are several number-worlds, as there are several Cultures. We find an Indian, an Arabian, a Classical, a Western type of mathematical thought and, corresponding with each, a type of number—each type fundamentally peculiar and unique, an expression of a specific world-feeling . . . the soul of that particular Culture.

Spengler states, "The most valuable thing in the Classical [ancient Greek] mathematic is its proposition that number is the essence of all things *perceptible to the senses.*"

This devotion to the corporeal "here" and "now" made "*zero as a number*" an impossible conception. "No conception of zero as a number could possibly come," Spengler says, "for from the point of view of a draughtsman it is meaningless."

Reading this reminds me of Piaget's children who confused words, perceptions, and things, and said there can't be words for things that don't exist.

Spengler then adds a key point. The Greek rejection of zero was deliberate.

> Babylonian and Indian mathematics had long contained, as essential elements of *their* number-worlds, things [including zero] which the Classical number-feeling regarded as nonsense—and not from ignorance either, since many a Greek thinker was acquainted with them.

This shed helpful light on my inability to explain zero to applicants. It reminded me of Greek tenacity, of how a handful had stalled the Persian army at Thermopylae. Who could overcome the soul of Greek culture in a job interview?

The prominent American mathematician Cassius Jackson Keyser (1862–1947) expanded the theme in his essay *Mathematics as a Culture Clue*. Keyser pictures us as so convinced that Classical and Western mathematics are "an earlier and a later stage in one continuous development" that we can't imagine them as "the offspring of two mutually alien Culture souls." Yet it is a hypothesis Keyser feels "we are bound to entertain."

And in the entertaining, Keyser finds more evidence for than against the hypothesis. He notes that the Greeks so differed from us in their mathematics that they would have regarded some modern assertions—for example, Henri Poincaré's description of mathematics as "the science of the infinite"—as "utterly incomprehensible, if not insane."

The Greeks tied numbers to sense perceptions of foreground objects, to bounded finite things. They didn't think in terms of empty extended space; they had no word for it. They thought in terms of shape and location. They eschewed zero, infinity, and change. They concentrated on the observable, the small, the unvarying. And so, as Keyser makes clear, they were stuck.

> A Culture whose mathematics is, like Euclidean mathematics, *functionless*, knowing nothing of energy, nothing of time or motion or velocities and accelerations, will have a Physics which, like that of the Greeks and Romans, is nothing but Statics. But a Culture whose mathematics is . . . almost exclusively concerned with *functional* relationship, will inevitably have a Physics which, like that of the West, is almost wholly Dynamics.

How, then, did we acquire zero, this number at once so vital for western mathematics and so alien to the Greeks?

Robert Logan, associate professor of physics at the University of Toronto, supplied the answer in an article titled "The Mystery of the Discovery of Zero."

We first encounter zero in primary school, Greek geometry in secondary school, and Greek logic in college. Many people therefore assume, says Logan, that zero was a less sophisticated part of *Greek* math. Yet Greek "mathematics was completely devoid" of the concept of zero. He then summarizes the discovery of zero and its astonishing mathematical consequences.

> Zero was an invention of the Hindu mathematicians, working more than 2000 years ago. Their discovery of zero led them to positional numbers, simpler arithmetic calculations, negative numbers, algebra with a symbolic notation, the idea of infinitesimals, infinity, fractions, and irrational numbers.

Ancient commerce, according to Logan, likely made both Greek and Hindu mathematicians familiar with the use by Babylonian scribes of zero as a place holder, much as the zeros in 2001 are place holders to tell us there are no hundreds and no tens between the two thousands and the one one. The Babylonians, however, did not develop zero in its own right. Our applicants who used zero in writing large numbers but wouldn't use it alone honored this precedent.

"The great mystery of zero," Logan quotes Constance Reid's *From Zero to Infinity*, "is that it escaped even the Greeks." If both the Greeks and the Hindus knew of the place-holding zero, what caused the Greeks to miss its potential for development?

Overzealous logical rigor, that's what. Beginning with the formal proofs of Thales, the Greeks elevated logic to the highest intellectual status. That led, says Logan, to a crucial argument by the philosopher Parmenides that "non-being could not *be*, because it was a logical impossibility." American pragmatist Charles Sanders Peirce quotes Parmenides, "Being only is and nothing is altogether not." Either way, note the concept of non-existence (addressed by Piaget) and the verb "to be" (addressed by Korzybski) again causing trouble.

Change was also impossible; for if the state of a thing changed, its original state would not be. Because non-being was impossible, change was impossible. These arguments were accepted. Logic is logic. Both change and non-being were rejected. This made zero, non-being, unacceptable.

Logan notes that the Hebrews agreed with the Greeks in regarding the idea of non-being as dreadful. The implication for Christian tradition is obvious.

The situation in India, writes Logan, was different.

> For both the Hindu and the Buddhist, the notion of non-Being was a state that they actively sought in their attempt to achieve Nirvana, or oneness with the whole cosmos. Non-Being was *something*—a state that could be discussed.

Just as Greek logic worked against accepting zero, Hindu and Buddhist philosophy encouraged it.

The development of zero began in India before 200 B.C. The Hindu symbol for zero, and for the unknown, was an oval enclosing an empty space. This makes visual sense; the Mayans quite independently later used a similar symbol as a place holder. Our zero evolved from the Hindu version.

The Hindu name for zero was *sûnya* (pronounced shunya), which meant "empty space" or "blank." By 1000 A.D. the mathematicians of Baghdad had appropriated the Hindu system. *Sûnya* became the Arabic *sifr* ("cipher"), which also means "empty space." Later, in Europe, Logan says, local opposition forced the new system underground. Hence "cipher" came to mean "code," and "zero" (from *zephirum*, a Latinized form of *cipher*) took over.

The new Hindu mathematics using zero was passed on to Europe through the works of the ninth-century Arab mathematician Al Kworismi. His name survives in our term for a mathematical procedure, "algorithm." His book, *On Restitution and Adjustment*, in Arabic, *Algebar wal Muquabalah*, gave us our term "algebra."

This recitation, which I have merely assembled from what others have said, should demonstrate that the math our schools teach has at least two vastly different roots. The Greek root gets much praise. The Hindu root via Baghdad gets little.

Yet western mathematics owes its eventual progress beyond the Greeks to its receipt in the Middle Ages of Hindu notions from Arab scholars.

How does it happen, then, that late in the twentieth century our quiz found schooled westerners keeping to the ancient Greek idea that the notion of nothing is self-contradictory, refusing to accept zero even as a point on a scale between, for example, somebody owing you a dollar and your owing them a dollar?

What we probably have here is our longest-running and best-documented cultural mathsemantic problem.

So, if zero-trouble has survived the last twenty-five centuries, what chance have we of licking it in the next decade? Forget that "nation-that-could-put-a-man-on-the-moon" routine. That was math-science-technology for a few. This is mathsemantics for everybody.

If you think math teachers could do it, given the tools, read Gina Kolata's article in the *New York Times* of April 2, 1980, about a teacher still promoting the old Greek view.

> When I was in school, one of my teachers told my mother that she was having the worst time trying to teach her class that zero is not a number, a notion that could only strike a mathematician as bizarre.

I do not mean to imply that math teachers teach poorly. On the contrary, math teachers apparently often succeed at teaching what they set out to teach.

Take algebra. According to the *Philadelphia Inquirer* of February 11, 1990, "educators are focusing on math because of widespread evidence that the taking of algebra is the single best predictor of whether a student will attend college." Perhaps this explains why most of our job applicants aced the test's two algebra problems.

$$\frac{x}{2} = \frac{3}{6}, \quad x = \underline{\hspace{1cm}} \qquad\qquad \frac{20}{x} = \frac{5}{2}, \quad x = \underline{\hspace{1cm}}$$

One hundred thirty applicants, or two out of three, gave the correct answer ("1") to the first algebra problem. One hundred eighteen applicants, or three out of five, gave the correct answer ("8") to the second. Al Kworismi would be proud. Too bad only forty-eight knew zero was the nearest whole number to .098.

Once aware of zero-problems, you'll find lots of them. One thing that opened my eyes was trying while on the witness stand to decipher what our analysts and clerks had meant by worksheet zeros, dashes, blanks, NAs, and unrounded entries.

In the course of standardizing our procedures, we found people who couldn't stand to throw anything away. If faced with, for example, ".098" in a table to be rounded to whole numbers, they'd post a dash, a "1," or ".1"; anything but "0." I even considered having our office use the chemical analyst's term, "trace," but decided it didn't get to the root of the matter and wouldn't help win cases.

Zero-problems come in enough guises to confuse anybody. We probably will never achieve total agreement.

Cassius Keyser criticized Korzybski for using "non-Aristotelian" (abbreviated $\bar{A}$) improperly. Harold Drake, a general semanticist, criticized Donald Wollheim, an editor at Ace Paperback, for error through typographical penury in changing the science-fiction title *World of $\bar{A}$* to *World of Null-A*.

If that's too obscure, how about "first floor"? This sometimes means the street level ("zero" floor, ground floor) and sometimes, as in French, the next floor up.

Speaking of French, the word *personne* can mean either "anyone" or "no one," depending on context.

"We have one-year-olds but no zero-year-olds," penciled math-department-head Edward F. Gardner in the margin while reviewing an early draft of this chapter. "This confused the hell out of me when I was five and my brother was not yet one."

Then there's "I *couldn't* care less," which logically means "I care so little there's no way to care less." Yet people seem, illogically to me, to mean the same thing when they say, "I could *care* less."

Stuart Mayper, editor of the annual *General Semantics Bulletin*, criticized the ".098" in my quiz because the best scientific practice is to begin with a zero ("0.098") to emphasize the decimal point that follows. Okay, I see that. My daughter in Texas, in defense of the quiz, bless her, criticized the criticism, "Why? To give people a clue? Good grief."

The aforementioned Gina Kolata said that one consequence of the "failed experiment in education called the new math" was instilling in pupils the "peculiar habit" of using the symbol for the empty set, "Ø," as a "strange way to write zero." She added that this distorts set theory. No argument, but please tell the designers of some of my computer typefaces.

How about "inaction"? Is taking no action an action? Is taking no decision a decision?

My daughter in California sent along a Don Addis "Bent Offerings" cartoon. A male shopper stands dumbfounded, staring at his "take-a-number" zero ticket.

People really can get worked up over zero. Ann Landers reported more than five thousand readers told her she had "batted zero" in siding with a reader who favored "zero" over "oh." How, one asked, would the song sound with the refrain "Pennsylvania Six Five Zero Zero Zero."

In a first-year physics lab experiment at college, I was one of many who got the wrong answer simply by counting "one" rather than "zero" on releasing the pendulum for its first swing in a timed series.

What can we do? Having everyone take up Hinduism or Buddhism to validate nothingness seems impractical. We do need something, however, to get youngsters off on the right track, to show them that zero is a number.

How about:

1. Put a "0" at the left end of every ruler. Don't say there's no room. Look at the "12" at the right end. This requires action only by ruler manufacturers.

2. Start counts with zero, in print and orally. Zero, one, two, three, etc. Don't say it's confusing. Conductors and starters give preliminary beats.
3. Tell math teachers school policy says zero is a number.
4. Hang a large dummy thermometer with a prominent zero and a movable degrees pointer in each kindergarten.

I'm sure you can think of more.

Proposition 10: Nothing is a mathsemantic problem.

Divorce

Divorce can have unexpected consequences.

In late summer 1944 a letter from home found me in an underground command post in France. It contained a clipping from the *Philadelphia Inquirer* telling of Mother's remarriage.

"Look," I said aloud, recovering from my surprise and proudly holding up the story, "my mother's gotten married."

One soldier slowly sized me up, head to toe and back. "Well," he drawled, "I'd say it's about time."

My wife was only an infant when her parents divorced. Her mother remarried once, her father three times. One result was that, counting my father's lady friend, our own children ended up with eight grandparents, exactly twice the usual four, and plenty of presents under the tree.

I remember once thinking I'd called my father long distance, but Mother answered the phone. This was thirty years after they'd separated and twenty years after Mother had remarried. I figured I'd dialed wrong, invented an excuse to get off the phone, and then called Dad again. Mother answered the phone. That may have been the only time she visited him in thirty years. I don't know. I was raised not to ask personal questions.

When you've gotten used to thinking of two people as divorced, or two things as separate, you take the separation for granted. It's imagining them together that's difficult. We run into this problem when we think about math or language.

A German correspondent tells me that J. W. Gibbs, author of *Elementary Principles of Statistical Mechanics*, during a faculty discussion at Yale on the relative merits of language and mathematics, is reported to have said, "Mathematics *is* a language." Apparently a reminder was needed.

We have this odd situation: We say mathematics is a language, but in practice we divorce the two fields. Listen to John B. Slaughter, chancellor of the University of Maryland.

> Eleanor Wilson Orr's book makes a major contribution toward our understanding of the ways in which language differences can affect the performance of black students in fields [mathematics and science] that do not seem to be closely connected to language skills.

Orr describes herself as "a high school teacher of mathematics and science" who "began to realize that many of the difficulties my students were having were rooted in language." Her book, *Twice As Less: Black English and the Performance of Black Students in Mathematics and Science*, should leave no doubt that "language can be a barrier to success in mathematics and science."

Elisabeth Ruedy, math teacher and the author with Sue Nirenberg of *Where Do I Put the Decimal Point? How to Conquer Math Anxiety and Increase Your Facility with Numbers*, advises attacking math problems by careful reading in small segments. "I have found," she reports, "that *misreading and misinterpreting* the facts lead to at least half of the errors people make with math problems."

The English mathematician George Boole (1815–1864), however, didn't overlook the connections between math, logic, and language. He made them the basis for modern symbolic logic.

I purpose to establish the Calculus of Logic, and . . . claim for it a place among the acknowledged forms of Mathematical Analysis. . . .

That which renders Logic possible, is the existence in our minds of general notions,— our ability to conceive of a class, and to designate its individual members by a common name. The theory of Logic is thus intimately connected with that of Language. . . .

. . . The mathematics we have to construct are the mathematics of the human intellect.

What kind of a language is mathematics? The question really should be asked and its answers discussed. They clarify both mathematics and language, both spoken and written.

Our mathematics, like Chinese but unlike English, is an ideographic language. English words are phonograms, indicators of pronunciation. Chinese characters are ideograms, indicators of meaning. Ernst Mach (1838–1916), Austrian physicist and philosopher who influenced Einstein, put it this way:

In Chinese writing, we have an actual example of a true ideographic language, pronounced diversely in different provinces, yet everywhere carrying the same meaning. Were the system and its signs only of a simpler character, the use of Chinese writing might become universal.

Mathematics has simpler signs and is well-nigh universal. Its written characters and symbols, 0, 1, 2, 3, 4, . . . , 9, +, −, =, and %, etc., are immediately intelligible to adults who share no common speech. Differences in written conventions, like plain sevens versus crossed sevens, or the switched meanings of commas and periods, "$1.300,00" versus "$1,300.00," are minor.

Mach pointed out that children (presumably German-speaking, but English-speaking would also qualify) can read words they do not understand, whereas Chinese cannot. Coming upon a new symbol, Chinese know neither what it means nor how it is said. Math is like that. Open a math text in a field strange to you. You can't read it. You know neither what the symbols mean nor what they are called.

I'm not qualified to pursue the point, but I wonder whether just being familiar with ideographic languages might make math easier. If so, wouldn't East Asian students get an advantage?

Constance Kazuko Kamii, in her excellent book *Young Children Reinvent Arithmetic: Implications of Piaget's Theory*, says in passing, "It is unfortunate that in English the counting (spoken) words from eleven through nineteen correspond so poorly with the Arabic written system." That is, eleven doesn't say one-one, twelve doesn't say one-two, thirteen puts the three (thir) part before the ten (teen) part, backwards from the written Arabic numeral (13), and the reversal continues through nineteen.

When I learned to count in French, I found a different pattern. Eleven through sixteen (*onze, douze, treize, quatorze, quinze, seize*) sounded like altered one through six (*un, deux, trois, quatre, cinq, six*), seventeen was simply ten-seven (*dix-sept*), and eighteen and nineteen continued the same pattern. The ten was obvious, not distorted as "teen" is. Yet the French system seemed unnatural to me. I guess I regarded all the numbers up to twenty as independent. The French way of saying ninety-nine, "*quatre-vingt-dix-neuf,*" which translates as "four-twenty-ten-nine," seemed almost laughable. I'm sure it's natural enough in French. If you're translating, you have to multiply the four times the twenty and then add the ten plus the nine.

German and English agree from eleven to nineteen (*elf, zwölf, dreizehn, vierzehn, . . . neunzehn*), but then with admirable consistency German continues the reversal Kamii noted. Thus, twenty-one is *ein und zwanzig*, twenty-two is *zwei und zwanzig*, and ninety-nine is *neun und neunzig*.

Our system follows German word order through nineteen and French word order after that. Blame it on the wanderings of the Anglo-Saxons and the Normans. If our system makes a difference, we should be able to measure it. I'm not sure it does.

Orr argues that black English vernacular hinders math comprehension. Prepositions cause real misunderstandings. Yet, despite widespread interest in both black English and math, the usual math-language divorce prevails. As Orr puts it:

> Linguists have identified features of black English vernacular (BEV) that can interfere with a BEV-speaking child's learning to read standard English, but no study has been done of the features of BEV that can interfere with the speaker's learning mathematics and science.

When Orr first suspected that her students' math difficulties related to language, she turned to linguists for help, only to have difficulty communicating with *them*. They presumed erroneously that her quest reflected a biased view of BEV as an inferior language. That view would have conflicted with a resolution of the Linguistic Society of America that all languages are equally capable of expressing logical thought.

However that may be, Orr's students confuse location with distance, and this confusion seems to be related to their unfamiliarity with such constructions as "from ... to" and "between ... and." They speak of "the distance between Washington to Cleveland." In subtracting, they take the smaller quantity from the larger regardless of directional indications, so that both 8 − 3 and 3 − 8 equal 5. They use "from," "minus," "subtracted from," and even "subtracted by," without regard to direction, so that agreement with standard math usage is accidental. Similarly, they treat "less than" and "less" as equivalent, although standard usage dictates a change of direction, so that "five less than ten" is five, while "five less ten" is minus five. They also treat division as directionless, so that "half" becomes confused with "twice," "2/3" becomes confused with "3/2," and "divided into" becomes confused with "divided by."

I presume most students have difficulty learning some math relationships and algebraic notations. I certainly did. Math can be work. I particularly remember puzzling over similar-sounding phrases that could have opposite math meanings, such as "divided by," "divided into," and "divided into parts." "Six divided by two" is three. "Six divided into two" is one-third. "Six divided into two parts," however, is again three.

Solving such puzzles, getting them straight once and for all, meant scrutinizing meanings closely. I can remember gaining confidence whenever my understanding of the words, such as "divided into" and "divided by," worked in agreement with the math operations I was being taught.

I could easily have been stranded early if "into" and "by" had not been part of my regular lexicon. I'm not proud of it, but I remember giving up on school math problems when I couldn't put them into words that meant anything to me.

Orr states my experience precisely.

> The standard distinctive uses of the English prepositions can in mathematics and science be an aid, perhaps to one's perceiving—and more certainly to one's preserving—the distinctions these prepositions indicate.

Orr may well be right about BEV, but that isn't my point. My point is broader. It relates to math and language generally.

My point is that divorcing math from semantics hobbles math. To remove the hobble, we must get into meanings, what we mean by meanings, and how math gets meanings.

Complex *language acquisition* is innate. All children under normal conditions learn to talk in ways so deeply complicated that linguists haven't yet sorted them out. All that seems necessary is that other humans talk to them.

Sound *semantics acquisition*, however, is not innate. Piaget's work shows that childhood semantics confuses the word with the perception with the thing. Korzybski's work shows that adults continue more or less to suffer from this confusion.

Before achieving mathsemantic know-how, then, we all have some semantic hurdles to get over. But apparently, based on Piaget's work, we don't even see these hurdles clearly much before we're about eleven.

The math hurdles loom up sooner.

Complex *math acquisition* is not innate. The fundamental reason would seem to be that mathematical ability has had little evolutionary effect. It's easy to imagine repeated prehistoric circumstances where good eyes, legs, memory, and vocal cords helped our species survive long enough to reproduce. It's harder to imagine scenes where mathematical ability tipped the balance.

The experiments of Professor Otto Koehler (1889–1974) show that birds can distinguish quantitative differences presented either simultaneously (say, marks on a container) or successively (say, number of seeds permitted before a rebuke). The ability varies by species. Pigeons can count to five, but become "hopelessly upset by changes in experimental conditions during training," such as switches from counting dots to counting mealworms. Jackdaws (a kind of Eurasian blackbird) can count to six and handle switches easily. Ravens and parrots excel; they can count to seven.

Koehler refers to this counting as "unnamed" and "prolinguistic." Language is not involved.

Koehler tested humans with figures presented on a screen too briefly to be counted linguistically.

> It is remarkable that, when this is done with human beings, the limit of achievement is of the same order as that shown by birds. Thus few persons reached eight, and many, like pigeons, get no further than five. . . . As far as we can see there is no expectation that man can achieve better results than the birds provided named counting is excluded.

The reason we need a math language, then, is that without language we're mathematical bird brains. We need a math language to help us do more with what we've got.

G. K. Zipf, a Harvard linguist who contributed much to statistics, demonstrated our innate desire to get more from less in his remarkable 1949 book, *Human Behavior and the Principle of Least Effort*. The basic points Zipf makes are that we are tool users and that we utilize our tools to minimize the amount of work to be done.

Mach held, according to Newman, that "if man were immortal he would be foolish to devise time- and work-saving methods, since they would only add to the tedium of eternity." But, being mortal, we have developed the language of math so that we save effort. "Mathematics," wrote Mach,

> may be defined as the economy of counting. . . .
> The object of all arithmetical operations is to *save* direct numeration, by utilizing the results of our old operations of counting. Our endeavor is, having done a sum once, to preserve the answer for future use. . . . It may happen in this procedure that the results of operations are employed which were originally performed centuries ago.

Mach's economy not only works, it works fantastically. We keep saving the old results. We keep improving our methods.

Thus, we start out roughly on a par with pigeons and ravens in being able to recognize counts up to about six or seven. Language then permits us to improve our mathematical ability enormously and to pass improvements to others. This is true both of our species and of each of us individually.

As we go, we also keep revising, extending, and improving mathematical language. Here's how professors C. I. Lewis and C. H. Langford put it:

> Arithmetic, at first, lacked any more appropriate medium than that of ordinary language. Ancient Greek mathematicians had no symbol for zero and used letters of the alphabet for other numbers. As a result, it was impossible to state any general rule for division—to give only one example. Operations which any fourth-grade child can accomplish in the modern notation, taxed the finest mathematical minds of the

Age of Pericles. Had it not been for the adoption of the new and more versatile ideographic symbols, many branches of mathematics could never have been developed, because no human mind could grasp the essence of their operations in terms of the phonograms of ordinary language.

So, if doubters ask why divorcing math from language makes a difference, you might quote them proposition 11.

Proposition 11: Without language, math is for the birds.

All normal children go well beyond birds, certainly, and perhaps all have the ability to learn complicated math. But few do. More people today than ever before wonder why. I believe part of the answer lies in math's being a language that does not conform to ordinary childhood semantics.

Recall that children fail to distinguish language from events, seek to grasp things whole, and look to purposes for explanations. Such childhood semantics works passably well in dealing with immediate, corporeal, finite things and family relationships.

Math, however, requires a different kind of semantics. Math's meanings depend on an endless series of detailed nonpurposeful relationships internal to the language. To understand these meanings, one must grasp that numbers form their own system apart from any specific events, that they relate to each other according to that system in endless particularity, and that their meanings derive neither from their names nor from what the teacher wants.

The child who unknowingly attempts to apply ordinary childhood semantics to math gets a surprise. The attempt fails.

Worse yet, there's no way for the child to construct the needed mathsemantics from ordinary childhood semantics. Numbers (one, two, three, . . .) aren't "things" in the sense the child knows. Names like "plus," "minus," and "equals" don't reside in objects. "Two plus three make five" isn't an arbitrary rule like "No more talking once the lights are out." There are more such seeming rules than anyone could possibly remember. There's no way to grasp math whole, once and for all. And there's no way to sweet-talk math away.

Math is a language. But perhaps not to most children; for even the child who welcomes math as a great new collection of interesting names quickly finds things going awry. I presume that the child most confidently locked into ordinary language, therefore more likely a girl than a boy, is the one who finds this unexpected failure the most startling.

Proposition 12: Ordinary childhood semantics can't handle math.

CHAPTER

Writer's cramp

M y father gave numbers added meaning for me.
     One day he handed my brother and me (about age eight and six) a pad of wide ledger paper and challenged us to start writing numbers in sequence from one up. We could take turns, continue for as many days, weeks, or months as we liked, and he would pay us some amount (I guess a dollar would be right today) each time we got to a one followed only by zeros. We set up a card table and chairs. He showed us where to start and left.

Our first reward was easy; it came at ten.

1
2
3
4
5
6
7
8
9
10

We sped on into double-digit numbers. I don't remember which of us, my brother or I, went first, but I remember waiting impatiently for my turn.

We quickly advanced through the twenties and thirties. We reached forty with the bottom of the first column.

"What do we do now?" we asked Dad. He showed us how to move right one column and put the "41" there on the same line as the "1" in the column just completed. He pointed out the double lines separating the columns and how each column was subdivided into smaller columns in different yellow tints. "Stay out of the wider subcolumn at the right," he said. "That's just for pennies."

We returned to our room, in a pleasant corner of our fifth-floor Philadelphia apartment. Sun streamed in through the casement windows above the shelves holding our blocks, books, and toys. The stuffed animals on our twin beds stared at us. We kept at the numbers.

41
42
43
44
45
46
47
48
49
50

The zero in fifty earned us nothing, of course. Only all-zeros after ones paid off. Still, it was reassuring to see the zeros coming up over and over again, signaling progress.

I remember stumbling over some numerals more than others. Threes, with their curls, were harder than twos. I experimented with different ways to write fours. I tried to make fives with one stroke, but settled for two.

Eights were fun, intersecting loops, but they were also unfamiliar, the easiest to overlook. I kept skipping them. Sometimes I didn't notice until reaching the bottom of a column that didn't come out evenly

to a zero. Then I'd have to go back, fix the column from wherever the
error had been made. Dad would expect good work.

74	
75	
76	
77	
79	78
80	79
81	80

Writing a whole set of eights was exciting. These were real numbers, I felt, because each was in its right place, in sequence where it
belonged.

Then a whole set of nines. So quickly, just a few minutes, including
seeing Dad about the columns, and here we were, coming to another
reward.

94
95
96
97
98
99
100

It was like collecting $200 each time you pass the "Go" corner in
Monopoly.

Speaking of Monopoly, we played it only a few times with Dad. He
introduced us to "auction" Monopoly. The person landing on an unsold property didn't automatically get to buy it as when we kids played.
No, the property went up for auction to the highest bidder. Dad was always complicating games that way. My brother and I would learn from
him, introduce the new wrinkle to our friends, and then have great fun
skunking them. They did the same for us.

We quickly exhausted the first page of ledger paper. How neat it
looked, from one to five hundred twenty. The first three columns on
the left, only one or two digits wide, looked thinner than the ones on
the right, which marched down the page three abreast.

We started page two.

521
522
523
524
525

Numbers ending in "5" were signposts. They seemed to say you
were halfway somewhere. Numbers ending in "25" also seemed to be
signposts. They were like quarters, big markers.

549
550
551
552
553
554
555

A triplet. I don't know when I first noticed them. Doublets hadn't
meant much. "11" was too familiar. We were going too fast to notice
"22" or "33." But "555" was like having three hands.

659
660
661
662
663
664
665
666

Sixes have quite a history. In St. Augustine's view

six is a perfect number in itself, and not because God created
all things in six days; rather the inverse is true, that God created all things in six days because this number is perfect, and
it would remain perfect, even if the work of the six days did
not exist.

Three sixes are something else again, the number of the Beast, as Umberto Eco's *Foucault's Pendulum* relates.

> [The Templars] were persecuted as heretics, and in their hatred of the Church they came to identify with the Antichrist. They knew that throughout the occult tradition 666 was the number of the Beast, and the six hundred and sixty-sixth year was the year of the Beast. Well, the six hundred and sixty-sixth year after 1344 [a date calculated earlier] is the year 2000, when the Templar's revenge will triumph.

I knew nothing of this, of course, but I'd already begun to look for whatever might be special in each number.

Some sandwiched different middles between matching ends: 707, 727, 737, 747, 757, 767. Boeing later used these as model numbers, starting with the four-engine 707. The company followed with the *three*-engine 727 and the *two*-engine 737. (Explain that to new clerks.)

I saw more patterns. Some numbers climbed internally, 789. Others descended, 876. Some descended in *double* jumps, 975, or *triple* ones, 963.

I loved nines. They heralded zeros, potential payoffs.

994

995

996

997

998

999

1000

The numbers suddenly looked more familiar. Why was that? Oh, I saw, it was because they now looked like dates, years, old ones to be sure, but still years.

1066

. . . .

1492

. . . .

1776

. . . .

1925 [the year I was born]

. . . .

1931 [the year I first counted this high]

By now we were on our fourth ledger sheet. Five hundred and twenty numbers per page (which Dad surely used in checking our work), in thirteen columns, each column allowing the counting to progress another forty.

Forty. There's an interesting number. Ali Baba and the *forty* thieves. Didn't it rain for *forty* days? Educator John Gummere (rhymes with Montgomery), in *Words &c*, based on his *Philadelphia Inquirer* language column, found forty turning up more than its share. "Moses was on the mountain forty days and forty nights. . . . Jesus 'fasted forty days and forty nights' (and so Lent lasts for forty days)." "We 'catch forty winks.' An old description of tousled hair was 'It stood up forty ways for Sunday.' . . . Before Buddhist burial services can be performed, the ashes of the deceased must wait for forty days. . . . The number of days for which patients with various diseases (e.g., scarlet fever) used to be isolated [was] forty days, and the very word *quarantine* is from Italian *quaranta*, 'forty.' "

Question: Could the naturalness of forty add mystery to *The 39 Steps*?

We encountered multiple zeros without rewards: 2000, 3000. And even more interesting patterns: 3456, 3579, 3663, 4774, 5432, 5555, 5678. I began to see the 5678 pattern not only in 6789 but also in 7890 and 8901. The digits 1 through 9 and 0 circled inexorably as if on one-digit-wide columnar wheels. Like a car's odometer.

Each columnar wheel to the left rotated more slowly. It had to wait for the wheel to its right to get to "9." The fourth wheel on the left waited for the third, which waited for the second, which waited for the first. No matter how fast we wrote to turn the first columnar wheel, the fourth seldom rotated.

8902
8903

. . . .

8909
8910
8911

. . . .

8918
8919
8920

. . . .

8929
8930

. . . .

8938
8939
8940

. . . .

. . . .

8988
8989
8990

. . . .

8997
8998
8999
9000

Still no payoff! Another thousand to go. Boring. Look for interesting patterns: 9009, 9090, 9119, 9229, 9630, 9753, 9876, 9889.

9898
9899
9900

Still a hundred to go! Oh well, . . . 9911 . . . 9922 . . . 9933 . . . 9966 . . . 9988 . . . 9990 . . .

9997

9998

9999

Hold it! The magic moment. All nines. A new column creeps up at the left, circling into existence.

10000

My brother and I no longer fought for our turns. But we were still fresh to the task, eager to see the new numbers.

Longer numbers brought more interesting patterns: 10001, 10101, 11111, 12221, 12321, 12345, 13579, 14141, 14703, 15551, 19991.

19997

19998

19999

20000

Dad had suggested we number the ledger pages. We were now on page thirty-nine. Each page for some time had looked much fuller than the first few. The columns now marched relentlessly down the pages, five numerals abreast.

I'd long ago stopped forgetting my eights. I'd learned to write threes patiently. I'd discovered that fives followed by ones can fool you into thinking they're fives followed by sevens. I'd learned that sharp pencils write more finely than blunt ones. I'd learned that paper sags in the middle when you write a lot of numbers on it. I'd learned that fingers can get tired of holding a pencil.

Yet new numbers beckoned upwards. Like Everest, they were just there. Some were lovely: 22222, 23456, 24680. Others were jagged: 31537, 35491. Some seemed like Mom's voice exercises: 46864, 57975. Others lacked immediately discernible pattern: 58921, 61309. Some seemed to blare like trumpets: 67799, 88599. Others wasted away: 97421, 98310.

With all, no two numbers were alike. Each was unique. Not "more" unique. Not "absolutely" unique. Just unique (from *unus*, one). Just one of each.

And now one that promised a reward circled up.

99998

99999

100000

Soon, long before two hundred thousand, my brother and I gave up. We could've gotten to a million, but it wasn't worth it, not for the dollar, not for the glory, and not for the further insights.

"If I counted one, two, three, and so on, one number every second," noodled Casaubon, in Eco's *Foucault's Pendulum*, over how to break a computer code, "it would still take me almost thirty-two years to get to one lousy little billion." (A billion, however, is a thousand times the million my brother and I let go.)

Fourteen pages later Casaubon observes:

> I believe that what we become depends on what our fathers teach us at odd moments, when they aren't trying to teach us. We are formed by little scraps of wisdom. When I was ten, I asked my parents to subscribe to a weekly magazine that was publishing comic-strip versions of the great classics of literature. My father, not because he was stingy, but because he was suspicious of comic strips, tried to beg off. "The purpose of this magazine," I pontificated, quoting the ad, "is to educate the reader in an entertaining way." "The purpose of your magazine," my father replied without looking up from his paper, "is the purpose of every magazine: to sell as many copies as it can."

I have no idea why Dad issued the ledger-paper challenge. Perhaps he wanted to ease Dick and me into accounting. He once, just once, made me a handsome offer, one I cherish, to become an accountant. But I turned accounting down as having the highest ratio of routine to exciting work of all the professions. Dad may just have wanted to keep us quiet and then, in his waste-not want-not style, have found some use in silence.

I estimate that my brother and I spent about thirty hours each on this lesson into the meaning of number, thousand, and million; into positional notation, the powers of ten, and infinity as a process; into

uniqueness, transformation, and recurring patterns; and into persever-
ance and the economics of diminishing returns to scale.

It's what general semanticists would call an extensional lesson, moving numbers around until you get a real feel for them. Until they mean
something.

Cheap, wouldn't you say?

My brother has since made a fortune as an electronic physicist and
I've gotten by as a market researcher.

I hope it doesn't bother you that my counting story is an adult recon-
struction of a childhood memory. The details don't really matter. What
matters is the great power of a lively extensional lesson.

Cassius Keyser once wrote, "The most difficult thing that teaching
has to do is to give a worthy sense of the meaning and scope of a great
idea." I take it that Arabic numerals, the decimal system, zero, and positional notation are all examples of great ideas. If so, the ledger-paper
challenge was the most meaningful mathematical lesson I ever received, even though none of those great ideas was mentioned.

My father also had my brother and me flip a coin exactly one thousand times and record each answer. Dad said he was interested in the
result, and perhaps he was. My brother and I took turns flipping and
recording.

In this extensional lesson we recorded the results of chance events
until we got a feel for them. That one afternoon's experience gave me
a continuing advantage over anyone who'd only read about successive
coin flips; for when I read years later that each possible sequence of
flips of a given length is equally likely (say head-head-head-head vs.
head-tail-head-tail), I understood it the same way I understood each
number was unique. I knew it because I felt it and believed it and because I'd been there. I'd done it.

My brother tells me what he remembers most clearly from the flipping was that Dad questioned whether our results were real. We can,
Dick discovered, find so many patterns and coincidences in random
events that they look unrandom, contrived.

Extensional lessons can, like ledger-paper challenges, give meaning
to pure math or, like coin-flips, to applied math.

When our own children were only about three and five, my wife and I taught them the semantic difference between one and two. We offered a reward for each pair of identical things they brought us. At first they offered grossly different things having the same name, like two pillows. We'd point out a plain difference, like color. They got smarter and brought two wooden blocks, then two pennies, then two pins, all without success. There was always some difference that they could see and accept. I marveled at their continued interest. They learned in two days and forever that one and one make two different things, that aggregates always involve differences, that all we ever add are apples and oranges. Even with this devastating information, both later did well in math.

Proposition 13: To make math your own, take vivid extensional lessons.

I recall many extensional mathsemantic lessons from my childhood, mostly learned while playing games. It seemed everybody played games. Dad played Hearts and Poker. Mother played Bridge. We kids played War, Go Fish, Concentration, Eights, I Doubt It, Fan-Tan, Rummy, and various kinds of Solitaire. We also played Tic-Tac-Toe until we realized it was totally predictable, that the second player can always force a draw. We also played Dominoes, Checkers, Parcheesi, and Backgammon. By the time I was twelve, we'd graduated to Bridge, Poker, and Doubles Chess.

In my teens, I played Bridge on Sunday afternoons with Comfort Avery Adams, Gordon McKay Professor Emeritus and former dean of the Harvard Engineering School. His son and nephew completed the foursome. Mrs. Adams lowered the window shades, so the neighbors wouldn't see us profaning the sabbath. I remember being told, "To play cards well is a sign of proper upbringing; to play cards too well is a sign of a misspent youth."

But there's nothing like playing games to prove that math and logic work. A seven beats a six every time. Two five-point chips always equal one ten-point chip. Taking eight tricks in a ten-trick contract is al-

ways "down two." There's only one correct total score. Disagree-
ments between players focus attention and sharpen wits. One learns
that the mathematical aspects of such disputes have necessary solu-
tions.

Kamii's choice of a title, *Young Children Reinvent Arithmetic*, reflects
her Piagetian training. We err, she says, in trying to teach children
math as if it were a set of social conventions or extensional facts. It's
not. It's a way of thinking about abstract systems of relationships.
Therefore, children must reinvent math in their own heads. She's convinced they would all do so and all come up with the same basic math
ideas if given the means.

Her book, partly coauthored by Georgia DeClark, a practicing first-
grade teacher, describes how they rewrote the math curriculum, got
children interested in numbers, kept them interested, let them go at
their own speed, got them to interact with their classmates about numbers, created no budding math phobias, made parents happy, and
turned out free-spirited students who handled math tests just as well
as those who'd drilled all year on workbooks, flash cards, and quizzes.
Can you guess how?

They gave the children an interesting collection of games and got
out of the way.

Proposition 14: Games can provide vivid extensional lessons.

There's nothing magical about the games Kamii and DeClark used.
Some are standard, like Tic-Tac-Toe and the card games War, Go Fish,
and Concentration. Others are variations of commercial games, like
Parcheesi and Sorry. Still others are homemade games using dice,
cards, chips, coins, and dominoes. Almost anything that makes counting into a game might do. Give children a big enough choice and
they'll find the games that interest them, the games they can understand, the games that provide a mathsemantic foundation that they're
prepared to absorb.

Unfortunately, the extensional lesson that occupies the students in playing number-driven games may annoy the teachers. DeClark acknowledges running a pedagogical gauntlet. For example, as she and Kamii watched children figure how far to move after casting dice,

> Connie and I observed even the higher-level children pointing to and counting each dot on both dice to know the total. We were so irritated by this behavior that we referred to it as "pecking." I was annoyed because, according to our October testing, these children already knew the sums. Connie tried to reassure me that the children would give up the pecking when they did not need to do it, but I was not satisfied. I felt the need to intervene, which I later realized made *me* feel better but did nothing for the children. We tried encouraging the children to use dice with numerals rather than those with dots. But when given the choice, they chose the dice with dots. I did not know then that most of the pecking would stop by March.

The children also pecked the spots on cards. A two and a three became "one, two," and "three, four, five." A three and a two became "one, two, three," and "four, five." Over and over. The same thing with a four and a five: "one, two, three, four," and "five, six, seven, eight, nine." It was as if the children were never going to stop pecking and just start adding, begin saying, "three and two make five," "five and four make nine."

Consider your own tolerance as an adult playing a game with a child. Could you endure six or seven months of listening to a child count off the obvious? I don't think I could. Well, maybe I could if I got it through my head that children have to be children. But nothing I've done so far makes me think I have that kind of endurance. I really admire patient parents and first-grade teachers. My daughters and I seldom played the usual games. I got my kicks by inventing new ones.

In their own good time, though, Kamii's pupils *convinced themselves* that the answers always came out the same, that they could *rely* on two spots and three spots adding up to five. It wasn't a "fact" they accepted because the teacher said so. It was something they absorbed totally and unself-consciously, much as they had earlier acquired ordinary language.

I wonder how many children receive enriching extensional lessons in numbers and mathematical relations. The lucky ones get them before starting school; from then on their chances decline.

Kamii and DeClark tell why.

> The lesson we learned was that it does not do any good to push children beyond the method that is best for them. In time, they gave up the pecking of their own accord, but it was *very* hard to sit back and wait for that glorious moment to come!

Yes, but *why* does 2 times 2 equal 4?

If everyone had to feel individually superior to the same thing, and it fell my way to choose the thing, I'd want each of us to be language's master.

Collectively, over the long haul, we people do change language. We invent new words, and we change the pronunciation and meanings of old ones.

At any particular moment, however, far too many of us, like the children Piaget studied, regard language as a natural system, immutably correct as it stands.

Perhaps you recall Humpty Dumpty's brave words on this point. He'd just told Alice his necktie was an *un*birthday present. On grasping his meaning, Alice remarked,

> "I like birthday presents best."
> "You don't know what you're talking about!" cried Humpty Dumpty. "How many days are there in a year?"
> "Three hundred and sixty-five," said Alice.
> "And how many birthdays have you?"
> "One."

"And if you take one from three hundred and sixty-five, what remains?"

"Three hundred and sixty-four, of course."

[After reviewing the math, Humpty Dumpty continued] "—and that shows that there are three hundred and sixty-four days when you might get unbirthday-presents—"

"Certainly," said Alice.

"And only *one* for birthday presents, you know. There's glory for you!"

"I don't know what you mean by 'glory,' " Alice said.

Humpty Dumpty smiled contemptuously. "Of course, you don't—till I tell you. I meant 'there's a nice knock-down argument for you!' "

"But 'glory' doesn't mean 'a nice knock-down argument,' " Alice objected.

"When *I* use a word," Humpty Dumpty said in rather a scornful tone, "it means just what I choose it to mean—neither more nor less."

"The question is," said Alice, "whether you *can* make words mean so many different things."

"The question is," said Humpty Dumpty, "which is to be master—that's all."

Mastery was a recurring question at our house.

My strongest childhood dinner-table memories are of my parents bickering over something called the "Doc Hay diet"; a disagreement with my sister that escalated into the "famous chocolate bread-pudding fight"; an ice-cream cake packed in dry ice that totally resisted apportionment at my brother's birthday party; and the repeated fetchings of books, especially dictionaries and the *Encyclopaedia Britannica*, to illuminate questions. You might say we were a family of caring, if disputatious, searchers.

We respected books without kowtowing. My brother developed a reputation for repeatedly using the remark "The book's wrong." Just appearing in print didn't make statements right. I caught his attitude early and felt its liberating power. You may have noticed from my examples that I haven't outgrown it. Say it aloud often enough, "The book's wrong, the book's wrong," and you can learn to use it without intending any disrespect.

I've frequently advised clients to fire away at my findings and recommendations. "If we discover *now* that I'm wrong," I argue, "we'll both be better off." I really believe that. And, yes, of course you should apply it to this book.

Facing the majesty of the language called "mathematics," though, how does one develop a feeling of mastery? Can we possibly learn it all? No. That's not the road to mastery, anyway. Humpty Dumpty didn't memorize dictionaries.

I don't use mastery to mean knowing more math or more languages than the next person. That's like being an expert. When I say "mastery" I mean "glory." I mean having the good feeling that you're in charge.

Lots of people, perhaps ultimately most of us, come to feel some kind of mastery over ordinary language. We see we have a choice of words, that it's up to us. We give words meanings shared only with intimates.

But how does one master this language called math?

My brother and I at about age ten and eight started counting in the duodecimal system. We didn't know that's what it was called. We just starting doing it. It's a way of counting with twelve digits instead of the usual ten (zero through nine). The twelve digits we used were called: (if you want the mastery effect, from here on count along with me by saying all the boldfaced words aloud) "**zero, one, two, three, four, five, six, seven, eight, nine, zip,** and **zap.**"

I don't remember what prompted us. Perhaps Bing Crosby's radio show. Bing's occasional guest Victor Borge made up languages. One, I believe, we then called the "upper" language. Whenever you used a word sounding like a number, you had to move it up by one. "*Once* upon a time," became "*Twice* upon a time." "A ca*nine* *too*th" became "A ca*ten* *three*th." Cheating slightly, "Intes*tinal* *fortitu*de" became "Intes*eleven*al *fivetithree*d."

Or the immediate stimulus may have been a Ripley *Believe It or Not* funny-page feature showing members of a predominantly hexadactylic family dangling their six-fingered hands. That could have made us wonder how they counted.

Anyway, we started exploring our zip-zap language. We found that the number following "**nine, zip, zap**" had to be written with a one and a zero, because we'd used up all the single numbers and had to start using two of them together. Think of a car odometer rotating the digits into view, going from 000008 to 000009 to 00000zip to 00000zap to 000010. Once gone, a "0" on any wheel will turn up again right after its zip and zap.

Without thinking about it, my brother and I automatically called the one-zero by its usual name, "**ten**." This one-zero appeared as far along in the count, however, as the old one-two. Fascinating. Would this mystery lead to mastery?

Like Jack with his stalk, we climbed our bean-counter. "**Ten, eleven, twelve, thirteen, fourteen, fifteen, sixteen, seventeen, eighteen, nineteen,** uh, **zipteen, zapteen,** uh, **twenty**." We perched momentarily on the ledge at twenty.

Then upward, to thirty, forty, fifty, sixty, seventy, eighty. We started going faster. So far, each ascent to the next perch followed the same pattern. "**Eighty, eighty-one, eighty-two, eighty-three, eighty-four, eighty-five, eighty-six, eighty-seven, eighty-eight, eighty-nine, eighty-zip, eighty-zap, ninety**." We were climbing numbers in a new way, and it worked!

Then, suddenly, on surmounting the nineties, something scary. "**Ninety-six, ninety-seven, ninety-eight, ninety-*nine***," a momentary pause, "**ninety-zip, ninety-zap, . . .**" Then a blank wall. Where's the next toehold?

Omigosh. "**ZIPTY!**" *Of course.* After the *nine* of the *nine*ties comes the *zip* of the *zip*ties.

Sucking in breath, we scrambled up. "**Zipty, zipty-one, zipty-two, zipty-three, zipty-four, zipty-five, zipty-six, zipty-seven, zipty-eight, zipty-nine,** uh," careful now, "**zipty-*zip*, zipty-*zap*, zapty**.

Zapty-one, zapty-two, zapty-three, zapty-four, zapty-five, zapty-six, zapty-seven, zapty-eight, zapty-*nine*, zapty-*zip*, zapty-*zap*," oh glory!, *"one hundred!"*

Where on earth, no, where in multiplication-table-land were we? This "one hundred," this one-zero-zero, was not the familiar one. This "one hundred" was up twelve flights of twelve steps. A dozen dozens. It was the old "one hundred forty-four." A *gross!*

We repeated the exercise until we could do it without hesitation. "One, two, three, four, five, six, seven, eight, nine, zip, zap, ten." We grooved the hard spots, "zapty-nine, zapty-zip, zapty-zap, one hundred."

Then we tried some simple arithmetic. Two plus two was four. No problem. Count it off, "one, two," then another two places, "three, four." Three plus three was six. Four plus four was eight. But five plus five was zip. That's right. Zip. Count it off, "one, two, three, four, five," then another five, "six, seven, eight, nine, zip."

Two times two was still four. What is multiplication, anyway, but repetitive addition? Two times two means take two twice. Same as two plus two. Same as counting up to two, then counting two more beyond that.

Two times three means take three twice, add three to three. Or take two thrice, add two to two to two. Either way, same answer, six. We could check our little multiplications by counting them off as repetitive additions.

Two times four (count off four twice) was eight. Right. Two times five was zip. Right. Zip. Two times six was ten! Count off six twice. Ten? Right.

We eventually realized we needed single-digit symbols to represent zip and zap. We chose "t" and "e," unitary echoes of old "ten" and "eleven." Now, division. Ten divided by two was six. First group = "1,2,3,4,5,6" and second group = "7,8,9,t,e,10." Two groups with six in each. Right. Then ten divided by six must be two. Check it: "1&2, 3&4, 5&6, 7&8, 9&t, e&10." Six groups of two each. Right.

We kept on this way until it was becoming second nature. Eight plus two was zip. Eight plus three was zap. Ten divided by three was four. Three times five was thirteen. Thirty minus one was twenty-zap. Thirty divided by six was six.

We discovered that all numbers ending in an even number (0, 2, 4, 6, 8, or t [zip]) were still also even. Therefore, they could be divided by two.

Now, if you miss the spirit of this, you'll be tempted to convert each twelves-system (duodecimal-system) number back into the tens (decimal) system, do your calculation, and then reconvert to the twelves system.

But that's the hard way. It's easier just to learn a little twelves system. If you've ever looked at a full egg carton, you know the fundamentals of multiplication and division. Let a carton equal "ten." That's one-zero, one carton and no eggs left over. Then ten (a full carton) divided by two is six, ten divided by three is four, ten divided by four is three, and ten divided by six is two.

You'd think you could also do this with twelve-inch rulers and digital twelve-hour clocks. But they're hybrids with numbers printed in the decimal system. You'd need to change their "10" to "t," "11" to "e," and "12" to "10." Egg cartons work better.

The problem we all have here is breaking the identification of the word "ten," and the number "10" with "///// ///// many" at the object level.

My brother, who's both older and smarter than I am, after getting me started, withdrew as usual into his own musings. I may have played with zip-zap for a week before getting bored with it. Then I looked around for more.

People say things come in threes. Okay. Let's count in threes. All we've got to count with, then, are 0, 1, and 2. As soon as we count, "**zero, one, two,**" we've used a full rotation of the first odometer wheel; it hooks the wheel at its left and they ratchet up together: one-zero. "**Ten.**" Right. "**Eleven, twelve, twenty, twenty-one, twenty-two,**" all full, hook, ratchet, "**one hundred, one hundred and one, one hundred and two, one hundred ten, one hundred eleven, one hundred twelve, one hundred twenty, one hundred twenty-**"

**one, one hundred twenty-two, two hundred, two hundred and
one, two hundred and two, two hundred ten, two hundred
eleven, two hundred twelve, two hundred twenty, two hundred
twenty-one, two hundred twenty-two,"** all full, hook, ratchet, **"one
thousand."**

Wow. So fast. One thousand already. Now, where are we in familiar
tens-system counting? Let's count it out.

0 1 2 10 11 12 20 21 22 100 101 102 110 111 112 120 121 122 200 201 202 210 211 212 220 221 222 1000

0 1 2 3 4 5 6 7 8   9 10 11 12 13 14 15 16 17 18 19 20 21 22 23 24 25 26   27

If you don't get this, you could go back to where zip-zap begins and
try again, out loud. You don't have to get it, but are you really going to
let an eight- and a ten-year-old beat you?

So, one thousand in the threes system (ternary system) gets you to
twenty-seven in the decimal system. One thousand is ten times ten
times ten (10^3). Twenty-seven is three times three times three (3^3).
Beautiful. I didn't know what it all meant, but it had some kind of internal order. Only later did I discover that 10^3 is 1,000 in any regular
positional-notation numbering system (one where each odometer
wheel has the same number of digits; some, like those in digital
clocks, don't).

It was now easy—for me, that is, I'm the one who lived through
this—to see that the same quantity would be 23 in the twelves system.

0 1 2 10 11 12 20 21 22 100 101 102 110 111 112 120 121 122 200 201 202 210 211 212 220 221 222 1000

0 1 2 3 4 5 6 7 8   9 10 11 12 13 14 15 16 17 18 19 20 21 22 23 24 25 26   27

0 1 2 3 4 5 6 7 8   9   t   e 10 11 12 13 14 15 16 17 18 19 1t 1e 20 21 22   23

Playing around further, I discovered that this many at the object
level, //// //// //// //// //// //, would be written differently in each
positional-notation system from the binary through the twenty-eights
system, when it first gets its own single-digit symbol.

Now *that* promotes mastery. I could, if I really wanted, count off this
one object-level quantity in twenty-seven different but systematically
related mathematical idioms.

Nothing at all like this happens with Roman numerals. Ten times ten
times ten, for example, would be written "X times X times X." (How
about "X × X × X" for something really confusing.) It equals "M." Just

"M." Remember, no zero. The zero (*sifr*, "cipher") came with Arabic numerals and positional notation. Roman numerals are a tens system lacking zero and fixed-position notation.

I wonder what it must have been like to participate in the change-over from Roman numerals to Arabic numerals. It's hard to imagine people choosing up sides, fighting, passing laws, and going underground over numbers, the way people do today over free speech and abortion rights.

Yet a struggle did take place at the nonelectronic pace of the Middle Ages. Real issues were at stake.

Arabic numerals, whose superiority seems so clear to us, were known in Europe by at least the year 1000. However, Roman numerals were protected by a strong rear-guard action that prevailed for seven hundred years.

Why so long? Why would cities, like Florence in 1299, pass laws against using Arabic numerals?

The answer was an understandable desire for security: Security against forgery.

Arabic numerals are easily altered, a 1 into a 7, a 3 into an 8, a 0 into a 6 or 9. Further, one thin stroke instantly enlarges any number tenfold plus one; say, 27 to 271.

Roman numerals offer greater protection. It's hard to alter one Roman numeral, I, V, X, L, C, D, or M, into another. Further, each numeral can occur only in certain positions relative to the others. Changing 27 into 271, for example, requires converting XXVII to CCLXXI. How would you conceal that?

About the only easy forgery is appending ones (I, II, or III) at the end, not very rewarding. To thwart even this, an extra twist may be used to mark the last character. Thus, 107 becomes CVIJ, virtually tamperproof. (You can still find this safety feature today on medical prescriptions, where for example, in lower-case Roman numerals, ij means two, and vj means six.)

Keep in mind that contracts, financial books, wills, and other records were formerly written by hand. Trade at any distance took weeks and offered great opportunity for deception. You couldn't check by telephone. Liability was personal and prison awaited debtors. So great commercial cities like Florence and anyone in business or with property had reason to insist on Roman numerals.

Even after Arabic notation took over there were still good numerical causes. Thomas Jefferson led the battle to throw out the pounds-shillings-pence system in favor of a decimal currency within the computational skills of all citizens. "Decimals had been studied by mathematicians for two hundred years," says history professor Patricia Cline Cohen, "but America was the first country to put them to practical use." Quite a man, Jefferson.

You never can tell how far a stray fact may lead. Our French teacher was ill one day, and our headmaster, John Gummere, known affectionately as "Chief," took over. I'd had Chief for Latin. But this day was special. He put French aside and told us about Indo-European. The stray fact that stuck in my head was that the Indian word "maharajah" is related to Latin "magnus rex," to French "roi majeur," and among others to English "major royalty." "Maha" means great (major, majority, majesty). "Rajah" means king (regal, royal).

So all these languages from India to Europe to America were related to one prototypical Indo-European language. If you knew any one of them plus the rules for switching between them (much as Spanish "dad" becomes French "té "and English "ty," as in sociedad, société, and society), you'd get a great head start. They exhibit what mathematicians call "invariance under transformation."

Our now-Californian daughter once brought home Jan, her fiancé, his parents, and his younger brother, Ian. Being a younger brother myself, I took an interest in him. Did he know that "Jan" and "Ian" were variants on the same name? No. His parents apparently didn't either. My bringing it up was just one more faux pas by dear old dad.

Consider: Jan, Ian, Ivan, Johann, Giovanni, Johannes, Hans, Juan, Jean, Sean, and Yanko, among others, mean "John." And Ivanov, Janowicz, Janosfi, Jantzen, Jenkins, Johanson, Hanson, Jones, Johnson, Jackson, McCain, McShane, and Gionopoulos, among others, all mean "John's son."

The world looks different when you can see invariances under transformation. Like 10^3 being 1,000 in any positional-notation system. Seeing that promotes a feeling of mathsemantic mastery.

One pleasant summer evening when I was about sixteen I was visiting with a Mrs. H. of my parents' generation and a lovely Miss M., too old for me, perhaps twenty-one. We were seated on Adirondack chairs, the kind with those wide wooden arms, under a crabapple tree. I remember embarrassing myself by continuing too long with some now-forgotten schoolboy mathematical comment. When I finally paused, Miss M. said, "Yes, but *why* does two times two equal four?"

That was a good question. It took me totally by surprise. It was so basic. I'd never thought about it. I tried to act smart, but I was too flustered.

I want another chance!

What I'd do with the chance is ask what the question means. "Is your question," I'd ask solicitously, "about the verbal level or about the object level?" Miss M. would probably ask politely what I meant. I can picture myself explaining.

"Well, if it's about the object level," I'd say, "then," and with that I'd put my forefinger to my lips to signal silence, so she'd have to watch. Then I'd use my finger to sign step "one" as in charades. Next I'd pick up two crabapples, óó, and put them on the arm of her chair. Then I'd sign step "two," place two more crabapples, óó, next to the first pair, and open my arms in the "there it is" sign. I'd repeat the demonstration until it became obvious that taking óó twice yields óóóó at the object level.

Then, aloud, with appealing authority. "We usually refer to this many [holding up óó] as 'two,' although that's pure convention. We could also call it 'ten,' as I'll show you in a moment. We also usually refer to this [now holding up óóóó] as 'four,' although that's also pure convention. We could also call it 'eleven,' or even 'one hundred,' as I'll show. But at the object level all we have is [and here I'd demonstrate again that óó taken twice yields óóóó]. It stays that way no matter how we talk about it.

"We need to keep that in mind because at the verbal level we can talk about this phenomenon in different ways.

"Let's first count as we normally do, using just ten digits, 0 through 9. Then we call this many [showing óó] 'two.' If we take this many twice, then we have this many [showing óóóó], which we call 'four.'

"But let's say we had a different counting system, one that used only four digits, 0, 1, 2, and 3, before starting over, as we do in counting people for tables of bridge, or cards into tricks. Then we'd count this many [showing óó and then óó] as one table (one) with nobody (zero) left over. We could write that as 'one-zero' and call it 'ten.' We would count, [showing óó] as 'one, two,' and the next [showing another óó] as 'three, ten.'

"And if we had used only three digits to count, as we do when counting feet into yards, then we'd count, 'zero, one, two, ten,' which means 'one yard, zero feet.' One more foot would make it 'eleven.' So we'd count [showing óó] as 'one, two,' and [showing another óó] as 'ten, eleven.'

"*And*, if we used only two digits, zero and one, to count, as computers do, then we'd call this many [showing óóóó] 'one hundred.' We'd count, 'one, ten, eleven, one hundred.'

"So, the question 'Why does two times two equal four?' has at least two quite different meanings. That's why I asked whether your question refers to the object level [showing óóóó] or the verbal level. That is, are you asking me why this [showing óó] and this [another óó] equal this [holding out óóóó] or are you asking me why we *usually call* that 'four'?"

Of course, by then it might have been so dark nobody could have seen what I was doing.

At least Miss M. wouldn't have said, as did several young ladies later, "Oh, you're the serious type." First of all, I was too young for her to care. Second, Mrs. H. was the serious type. (You were wondering what she had to do with this story, right?)

When Miss M. asked me in that long ago why two times two equals four, did she mean at the object level or the verbal level? Did she realize that "times two" means "taking twice"? I don't know. I never asked.

As I said, Miss M.'s question took me by surprise. I'd not considered it. Before I'd ever had to write "$2 \times 2 = 4$" in any quiz or workbook, I'd already felt its truth in my bones. Not because of flash cards or any teacher's say-so, but because it just had to be. Because the numbers and number reasoning used at home and in games with friends worked.

No wonder math in the early grades was a total snap. I found myself hardly listening to explanations of how to add, subtract, multiply, and divide. I already knew how. I received no remedial attention.

My brother and friends kept me ahead of the pack. I remember my brother's teaching me long division at home the same day he learned it at school. I thought it was neat and he was great.

It was several years before any math class challenged me with concepts I couldn't immediately grasp. By then, however, I knew I could think things out on my own. I was more fortunate than I knew. I'd been spared the ordeals of not knowing and not understanding, the ordeals of trying to fathom what the teacher wanted, the ordeals of trying to remember uncomprehended math "facts" and procedures.

Ruedy reports in *Where Do I Put the Decimal Point?* that she "grew up seeing math as a simple, down-to-earth endeavor" and that she was fascinated by and loved numbers from an early age. She doesn't think this love unusual. "I have never found," she says, "a little child who did not love numbers." Ruedy obviously means preschoolers.

For, Ruedy says, once into school she quickly found math anxiety, not in herself, but in her classmates. "My compassion for math-anxious friends was helped along by my second-grade teacher, who pulled girls several inches off the floor by their hair when they could not answer a math problem."

One result may be lifelong math confusion.

> I have found out some people truly have no idea that at-
> taching a zero to a number makes it ten times bigger. They
> see absolutely no way of explaining, when two numbers are
> added together, what the "carry" is all about. They can't un-
> derstand why we start adding with the rightmost position,
> and what the carries are.

Lucky me. If I ever had problems with zero or with carrying or with which column drives which, the ledger-paper challenge removed them. My feel for such things isn't just in my head, it's in my fingers, my hand, my arm, my eyes, and even somehow in my love for my brother and family.

I should have asked Miss M. what her first grade was like.

"When I walk around first grade classrooms while children are working on workbooks," writes Constance Kamii, "and stop to ask individual children how they got a particular answer, they typically react by reaching for the eraser and start erasing the answer, even if it is perfectly correct! . . . Already in first grade these children have lost confidence in their own ability to figure things out."

Why do children lose confidence? Kamii's answer is simple: We stifle their ability to figure things out. We try to feed them math as "facts" and procedures that they're not ready to learn.

Young children don't have empty minds ready to be filled. They have their own ways of looking at things.

Many mothers, for example, know that their children would rather have a small *full* glass than a large *half-full* glass. Arguing that both glasses contain the same amount of liquid doesn't stop the tears. You can even pour the milk back and forth to show you're not cheating your child. No good! Young children judge amount by height. They don't understand the conservation of matter. Your insistence on repeatedly altering the amount of milk—for that's how they see it—may just drive them crazy.

Their understanding of numbers is similarly nonadult. For example, children judge total number by overall extension. They say two parallel rows of nine beads have the same number only when their ends line up exactly. Spread out one row and they say it now has more beads. Squeeze it back too much and they say it now has "less" than the untouched row. Yes, they see you didn't add or remove any beads. They just don't grasp that moving beads around doesn't change their number.

You think that's too far out? Well, don't take *my* word for it. Consult Piaget's *The Child's Conception of Number*. Then *you* repeat one of his experiments with a five-year-old.

For example, drop a bead into a glass jar while the child drops a similar bead into a jar of a different shape. Repeat, always noting aloud that you're both dropping a bead at the same time. After five or six beads, ask the child which jar has more beads in it. The child will point to the narrower jar. Why? The beads rise higher there, and the child in this case, as with milk in a glass, judges quantity by height.

So, conservation of quantity isn't understood. One-to-one correspondence isn't understood. Number isn't understood. At least not the way adults understand it. To a child, "three" is like TV channel 3, the name of something.

What, then, can the symbols "3 + 2 = 5" mean to a young child? Not at all what they mean to us. And, as if requiring an adult's concept of number were not taxing enough, such an equation demands an understanding of writing, plus and equal signs, and the formula convention. Children grasp things whole. If the child isn't ready for *all this all at once*, the equation is just so much nonsense. Flash cards and workbooks won't bridge the gap. It's too great.

Of course, an attentive child shown "3 + 2 = _____" often enough will learn that Teacher wants to hear "five" as *the correct answer*. A bright child who's eager to please can memorize a great many math "facts" and procedures without understanding what makes them work.

Some of these children grow up and take jobs using math. They think they're "good at numbers."

Others, like Miss M., continue to ask why math works.

Still others, perhaps the largest group, find that making teachers happy is insufficient incentive for getting personally involved with unintelligible exercises.

And some develop full-blown math phobias.

Many bright and articulate children who don't "get" math in school certainly grow into competent adults. Their example and their number make math look optional, its lack acceptable.

For obvious reasons, few people hating math or doubting their math talent answered our "good at numbers" ad. So I have no quiz results on them. Yet I suspect many, math scars aside, have more number-sense than some of our quiz takers. If I'm right, then the burden the doubters bear is as much psychological as mathematical, perhaps more so. If you happen to be of that hardy breed, I admire you. I hope you enjoy my book and find it useful.

Educators know something is wrong with how math is taught. So do journalists and letters-to-the-editor writers. They just don't agree on what.

I read a lot about teaching math these days, but little of it touches me. Somehow, I just can't get worked up over such "issues" as whether schools should concentrate on reasoning or drill, whether they should allow students to use hand calculators, or what math courses they should teach. I don't deny these are practical questions. But why argue about them? Why not just let schools and teachers try out the approaches they honestly believe in and then tell us, if they can, what differences they make? If, as I suspect, impartial observers would find few clearly material differences, that would confirm my suspicion that these "issues" leapfrog past a more fundamental point.

Kamii's first-graders played games and learned math. I suppose we could argue whether their play was drill or reasoning, but I see no point to it, because those categories don't catch what really makes games work.

The games' magical ingredient is neither drill nor reasoning. It's *autonomy*. Without autonomy, games don't work. Kamii reports an early lesson:

> I tried to introduce card games such as War and Concentration in a first grade classroom in Chicago. The teacher of this class assumed that it was her responsibility to control, direct, organize, and police her pupils. She dutifully divided the class into small groups to play the games. But the groups quickly dissolved as fights broke out among the children, and what was to have been a rich activity degenerated into screaming, shouting children and torn cards.

Children must be taught to be autonomous, self-regulating; for, as children, we are naturally heteronymous, looking to others—mostly grown-ups—for guidance. Kamii says we find autonomy hard to learn.

Consider a child reacting to a teacher's handing out math worksheets
and calling for attention.

> From the child's point of view, math thus becomes "answer-
> ing the teacher's questions for which he, the teacher, has the
> answers," since it is the adult who issues the questions and
> marks the answers as right or wrong.

Unfortunately, teachers also find autonomy hard to teach.

> If a child writes $8 + 5 = 12$, most teachers will mark it as being
> wrong. Thus arithmetic becomes one more area in which
> truth and reason are confused with adult authority.

The child, says Kamii, is thereby discouraged from thinking autonomously. "Inferences about truth that should be made by the child
come ready made, instead, from the teacher's head." The more the
teacher tries to show the child how to do a math problem, to add or
subtract, to carry or borrow, or when to reverse signs, the more autonomy fades. "Imposition of adult algorithms serves only to reinforce
children's heteronomy and to hinder the development of their natural
ability to think."
    It's tough for a teacher to get out of the way, to transfer control to
children. However, says Kamii, Piaget argued that "children must do
their own thinking autonomously to construct logico-mathematical
knowledge . . . because this knowledge must be constructed from
within." And, tougher still, intellectual autonomy can't stand alone. "If
children are silenced in the social and moral realm, they will not feel
free to express their ideas in the intellectual realm, either."
    Allowing children to learn math their own way, what an extraordi-
nary idea! Kamii gives an example:

> A teacher divided her second-grade class of 28 children into
> six small groups . . . and put the following problem on the
> board:

$$107$$
$$+117$$

The first group said the answer was 2114 because 1 + 1 was 2, 0 + 1 was 1, and 7 + 7 was 14.

The second group said the answer was 214. They agreed with the first group, except to say you couldn't put the entire 14 where only one number should go, and they chose to use the 4.

The third group agreed with the second, except they chose to use only the 1. Their answer was 211.

The fourth group said this problem involved "carrying," and got 224 for their answer.

The last two groups had no new solution.

The class argued for forty-five minutes, but agreed only that there couldn't be four correct answers. The teacher, for the sake of moral and intellectual autonomy, did not intervene.

Yes, the children eventually found the right answer and a way to get right answers.

Let children reinvent arithmetic, says Kamii. "Since there is absolutely nothing arbitary in logico-mathematical knowledge, children are bound to find the truth if they argue long enough."

Wow!

Proposition 15: Until you work it out for yourself, two times two makes four only because the teacher says so.

CHAPTER

Impurities

My advocacy of mathsemantics doesn't require choosing between math as a system of reasoning (pure math) and as a way of describing events (applied math). We need both.

When I hear people speak of "pure math" or "just math," I assume they mean to emphasize math's formal side. They mean math as a system of reasoning that stresses definition, proceeds by rule, and exhibits its internal consistency. Even so, they still usually mean math as a system of reasoning applied to the world. They don't take it as far as some "pure" mathematicians do.

The eminent Englishman G. H. Hardy (1877–1947), according to James R. Newman,

> was a pure mathematician. The boundaries of this subject cannot be precisely defined, but for Hardy the word "pure" as applied to mathematics, had a clear, though negative, meaning. To qualify as pure, Hardy said, a mathematical topic had to be useless; if useless, it was not only pure, but beautiful.

"I have," Hardy wrote in *A Mathematician's Apology*, "never done anything 'useful.' " "Nonsense," says Newman, advances in pure math have ways of becoming useful, unexpectedly.

When I hear people speak of "applied math," I assume they mean to emphasize math's personal, business, social, and scientific use. They mean math as a way of addressing a practical or theoretical matter in a field other than math itself.

Both pure and applied math cover inexhaustible fields that permit many approaches. My own view is that we must learn some pure and some applied math.

On the pure side, I'm impressed with the mathematical way of thinking. It works in many places. It has much to offer. I wouldn't want to do without it. Yet math as a system of reasoning can be oversold. I don't accept, for example, that math provides *the* way to think. I have at least four reasons.

First, there's Kurt Gödel's proof that no mathematical system can ever be complete in itself, not even theoretically self-sufficient, but must always contain statements not provable within the system. That's enough to warn me off total reliance on a purely mathematical approach.

Second, I'm leery of pure math's enticing anyone into a hopeless quest for certainty. Yes, solving a puzzle mathematically, witnessing a mathematical proof coming together, can be inspiring. But expecting one's reasoning or life in general to follow those examples can be demoralizing.

Third, even scholars who find much to admire in mathematical reasoning now doubt that it provides the best model either of how we think or of how we should think.

George Lakoff, for example, author of *Women, Fire, and Dangerous Things: What Categories Reveal about the Mind*, believes that mathematical reasoning, the classical model, has led to an inaccurate view of human cognition. He asserts his view forcefully, then adds the following, apparently as balance.

Finally, the present study is by no means antimathematical. It is quite conceivable that a mathematics appropriate to the kind of cognitive systems discussed here could be developed. In fact, I am rather hopeful about this.

Mathematical reasoning, according to Lakoff, promotes the view that categories distinguish members from nonmembers on the basis of common attributes. Essentially, an item is fully in a category if it possesses the requisite attribute; otherwise, not. In the classical model, for example, every number is an equal member of the class "number."

Evidence has accumulated, however, that this classical model understates the rich complexity of the human mind. We can group things tightly or loosely. We can use common attributes or scenarios and clusters of associations not captured by the classical model. Thus, in a modern cognitive model, the single-digit numbers from two through nine are shown to be *better* examples, called *prototypes*, of the class "number" than are one, zero, minus fifty-three, eighteen million, and so on. Zero and the others are members, but less favored, less typical members.

For example, consider the following answers to the question "Were there a number of them?"

—"Yes, five."

—"Yes, but only two."

—"No, just one."

—"There weren't any."

These answers treat *two* and *five* as *numbers*, but deny that status to *one* and *zero*. Actually, *two* just makes it. In the case of *one*, the answer confirms that asking about *number* in this context makes sense, although it then denies that *one* is a number in that sense. However, in the case of *zero*, the answer shows less readiness to accept that *number* is relevant.

The classical-model response would be that of the editor who advised me, "What this really means is that 'number' in this sense is *plural*. Therefore, one could answer with any positive number greater than one, even a 'less favored' number like 'eighteen million.' Not a good example, I think."

The response is perfect, however, in showing how quickly we would set up a new category, "plural numbers," to include all integers greater than one, within which we could then treat all members equally. Mathematicians have thus created separate categories of numbers (n) called natural n, cardinal n, ordinal n, odd n, even n, whole n, integers, fractions, complex n, real n, imaginary n, positive n, negative n, rational n, irrational n, transfinite n, and transcendental n, within each one of which every member is regarded as equal.

The modern semantic approach, however, is that we use the term *number* to refer to many related but different things, some of which are nearer the central, prototypical (perhaps especially 3, 4, 5, 6, 7) meaning than others. Thus, instead of separate categories of absolute equals, we have related usages with shades of meaning. Less rigidity, more feely-interpretation.

You may recall from an earlier chapter that three-fourths of our job applicants failed to treat zero as the "nearest whole number." In my view, and I believe that of most mathematicians, these applicants were wrong. We must be ready to treat zero as a number.

But let's face it. Even those of us who do accept zero as a number should admit that it's not a prototypical number. Why otherwise, for example, would nineteenth-century mathematicians have declared division by zero an inadmissible procedure?

Prototypes tend to be at the level most easily grasped. Thus we learn the category "cat" before we learn either subordinate (Siamese, tiger, cheetah) or supraordinate ones (mammal, predator, animal). Yet the classical model treats all category levels evenhandedly.

Apples and oranges are at the level most easily grasped. McIntosh, Winesap, and Valencia are at a less obvious subordinate level. Fruit is at a less obvious supraordinate level.

Therefore, the pronouncement "You can't add apples and oranges" strikes with particular force. We'd react differently if the rule were stated, "You can't add McIntoshes and Winesaps." We'd reject it, because it would promote a level less easily grasped than the natural one, namely, apples. This is what happened when I told applicants they couldn't add "reddies" and "greenies." They told me that of course they could, "because they're both pencils." So actually we don't treat categories evenhandedly.

Perhaps no people ever tried to apply evenhanded logic to common-attribute categories more passionately than the ancient Greeks. As we saw in an earlier chapter, their logic led the Greeks to reject zero as a number, which Europe then had to import fifteen hundred years later from an alien tradition. I suspect we'll ultimately find it odd that we ever exalted Greek logic and categorization as a model for clear thinking when its best proponents so clearly proved its pitfalls.

My fourth reason for not accepting that math provides *the* way to think is my work. Aviation market research has required me to combine math with a modern cognitive approach that has repeatedly illustrated the limits of the usual mathematical outlook.

Travelers every day rapidly reach decisions on how well airline schedules suit them. Yet it turns out that measuring the quality of airline schedules between two places is an inordinately difficult mathematical problem. The traveler's brain does easily what most skilled analysts, in my experience, can't capture adequately in numbers.

Why? Because our judgments of airline schedules for our own travel derive from what Lakoff would call an idealized cognitive model, a travel scenario. That model consists of going when we want to go, getting there in the least time, with the fewest stops, on the best equipment, having fallback positions if part of the schedule goes awry, arriving when we want to arrive, having enough time at our destination, being able to leave when we want to leave, but flexibly, and getting home when we want, among other considerations.

How can an analyst measuring airline schedule quality account for a combination of such complex judgments made by different potential travelers?

Most analysts start by counting the number of flights scheduled. If that's the first thought that came into your head, you're a normal product of our mathematical training.

"Scheduled flights" fall into a convenient category. Unfortunately, counting them, while telling us something, starts us down the wrong track. The counts lead to absurdities and inconsistencies requiring ever more adjustments. It's like starting with the assumption that the sun and planets revolve around the earth. To account for observed variations from expected paths, one must introduce complications (epicycles, deferents, equants, excentrics, and the like, now almost forgotten) in vain attempts to salvage the original plan.

I've never met a traveler who judged the quality of airline schedules for an upcoming trip by counting flights. Such counts require decisions on which flights to count: just nonstops? or also single-plane one-stops? other single-plane flights? listed connecting schedules? unlisted connecting schedules? in both directions combined? or separately? Travelers don't bother with such questions.

Travelers care more about getting a good night's sleep. If in their judgment a flight leaves too early, they go the night before.

Travelers have a feel for potential delay. They can judge when relying on the very last flight risks too much, and choose an earlier, slower but surer, service via recurring connections at a gateway.

My office can train a clerk in about three days to reach exactly reproducible, scaled measurements of airline service quality using a system based on a general travel scenario. A computer service firm gave up attempting to reproduce our measurements electronically when two years of unfinished effort promised a method costing ten times as much per measurement as doing the work by hand. The computer analysts were so steeped in classical categories they found it difficult to handle our system.

Like Lakoff, I'm rather hopeful that a mathematics appropriate to the kinds of cognitive judgments people make will be developed. Such math would open doors to all kinds of useful measurements of quality and quickly develop worldwide applications. I'd expect it to show with embarrassing clarity that more of almost anything is not necessarily better. Meanwhile, life must go on.

Speaking of embarrassing clarity, we can generate a good example of the inadequacy of mathematical reasoning just by trying to classify and measure mathematical reasoning. We've heard much lately about our need to increase numeracy. We've probably begun to believe that people used to be better at numbers, that our educational system has failed, and even that we know what we're talking about. However, our ignorance about math runs deeper than may be comfortable to admit.

Patricia Cline Cohen, in her history, *A Calculating People: The Spread of Numeracy in Early America*, faces this matter with winning candor.

> Ironic as it sounds, it has not proved possible to construct a quantitative measure of numeracy rates, both because of the indeterminate definition of the term and the lack of systematic historical evidence.

In other words, we don't know whether we're becoming more or less numerate, because we're not sure what we mean by numeracy and we haven't got good data anyway on how well we did earlier.

Now, a classical definition of numeracy would establish a category with clear boundaries that would enable us to *measure* numeracy. Could boosters of mathematics find anything more useful? Is anyone better equipped than they to define and measure numeracy?

Then why haven't they done it?

Cohen reports that the term "numeracy," according to the 1976 supplement to the *Oxford English Dictionary*, was first used in a 1959 report on English education. The *OED* defines numeracy as "ability with or knowledge of numbers."

But what is that? Literacy can be defined narrowly as an ability to read and write. I presume we don't want to confine numeracy to handling written numbers.

Literacy can also be defined broadly to mean versed in literature or creative writing. I presume we don't want numeracy to mean versed in mathematical literature or theory.

So what do we mean by numeracy? Or innumeracy?

Fowler's Modern English Usage says the 1959 Committee on Education defined numeracy as "not only the ability to reason quantitatively but also some understanding of scientific method and some acquaintance with science," and adds, "Clearly there is need for such a word; whether this one has come to stay remains to be seen."

John Allen Paulos defines innumeracy, at the start of his best-seller by that name, as "an inability to deal comfortably with the fundamental notions of number and chance."

A. K. Dewdney, who seemed to inherit the *Scientific American* mathematical mantle formerly worn by Martin Gardner and Douglas R. Hofstadter, says he prefers the term "math abuse" to "innumeracy," because

> it has a wider scope. It includes errors that are not strictly numerical, and it also has a moral dimension. We abuse mathematics by failing to apply even the little we know of it to the false or questionable ideas that we encounter.

These quotes give no reason to dispute Cohen's contention that the definition of numeracy is indeterminate.

I must also credit her argument that we lack systematic historical evidence on numeracy. She proves as much in her own interesting accounts of the growth of numeracy in Great Britain, colonial America, and the United States through 1850. I was particularly struck by her footnote stating that the most valuable works on the history of teaching elementary mathematics remain ones dating from 1896 and earlier. I would have thought a later history would be useful for the current debates.

Therefore, much as I admire the pure side of mathematics, I'm not about to succumb to the idea that mathematicians know best or that the mathematical approach provides a superior model for reasoning about everything. Not, at least, until mathematics can measure itself.

On the applied side, I'm impressed with the interest in math that can be created through practical examples.

According to the *New York Times* of March 1, 1989, students at the Studio Elementary School on the Upper West Side prepare lunch in a class called "cooking" by the pupils and "kitchen science" by the faculty. Reading recipes, measuring ingredients, and watching cooking times all have a bearing on math, and apparently the students love it. But math limited to such practical exercises would be limited indeed, and I don't presume that the school so limits it.

Obviously, applied math can be oversold. It could turn into fun and games, with no more bearing on life than the daily crossword puzzle and cryptogram. I won't labor the point.

I do advocate teaching more about the fuzzy boundary between pure math and applied math, especially the fact that it's fuzzy. Indeed, most math probably falls into disputed territory.

Some deep thinkers would argue that pure math must relate to nothing but its own symbolic relationships. From this point of view, if you think of Euclidean geometry as the geometry of a plane, then it ceases to be pure math. Most people would find this definition too pure.

Other deep thinkers would argue that applied math must involve real data having the potential to falsify the calculations. Most people would find this definition too narrow.

Which leaves us with a problem classifying the following made-up algebra problem. *A man is three times as old as his son today and in ten years will be only twice as old as his son. How old is the man now?*

This isn't pure math by the pure definition, because it's about people and chronology. Therefore, it must be applied math.

But it isn't applied math by the applied definition, because the answer, 30, is knowable with certainty in advance. Therefore, it must be pure math.

From a pure-math perspective, utilizing a classical approach to categories, we're likely to feel we need to settle this definition question, that something's wrong if we don't. From a cognitive mathsemantic perspective, we find resolution unnecessary. We know that most boundaries are fuzzy and that we tend to become our most intolerant of fuzziness when around math.

We must do something truly novel. We must face up to teaching math (1) as a pure symbol-system, (2) as a symbol system we can apply to events at our own risk, and (3) as an adventure in crossing fuzzy boundaries between the pure and applied domains.

Pure math is a language concerned with the consistency of its own internal processes. My childhood counting, the ledger-paper challenge and zip-zap, which steeped me in the workings of decimal and positional notation, are examples of pure-math training.

Applying math to events has occupied much of my professional life. The addition of apples and oranges, which many people reject as impossible, mirrors practical problems I've faced. So do multiplying-travelers-by-hours and my adventures with the meanings of "passengers" and "trips."

The fuzzy boundary between pure and applied math involves a changing mix of concerns. To traverse that boundary, one must somehow blend a desire for internal consistency with unavoidable approximations and definitional uncertainties.

I take my survival on the witness stand as evidence of successful boundary crossings. Cross-examiners could challenge me on virtually anything I'd ever said. Math and my applications of it were always fair game. I might be testifying for Savannah and be asked to explain why I'd used a different statistical method in a nonlitigated report written ten years earlier in an entirely different context for Kansas City.

If expensive lawyers couldn't find a way to demolish my credibility, that suggests it's possible to function with joint concern for math's internal logic and math's application to events. We don't have to choose between them.

Why aren't more people comfortable with both aspects of math? This is a big question. My short answer: Distinguishing pure math from applied math and then making one's peace with each requires some ability in both math and semantics, neither ability is innate, and our culture doesn't encourage the combination.

I suggest we hang loose, accept, enjoy.

How far is up? The pure-math answer is "twice half the distance." I know it's a childhood riddle-joke, but a lot of math is really like that: $a = a$; $a = 2(a/2)$; p implies q if and only if p implies q.

The applied-math answer is that it depends on what up you're talking about. If it's up from earth, then you could consider that up runs out once you've escaped earth's gravity.

What's the next term in the series 14, 23, 34, 42, . . . ? Looking just at the math, we might see the progressive differences of +9, +11, and +8 as widening oscillations whose next four steps would be +12, +7, +13, and +6. We could, therefore, take the +12, add it to the 42, and answer "54." That's one of many possibilities. There really is no single pure-math solution. The answer sought, however, is "59." Applying the series to the event level in New York, we find that 59 is the next street where diagonal Broadway crosses an avenue.

How far can you run into a wood? The just-math answer is "halfway, then you're running out." The applied-math answer is "farther than into a grove, not so far as into a forest; perhaps between a dozen and a few hundred yards."

If we see math as limited to its own internal consistency and by Gödel's theorem mentioned earlier, then we reach the unexpected conclusion:

> **Proposition 16:** Mathematics is a limited and logically incomplete system of reasoning.

Therefore, much as I admire (from a great distance) the work of pure mathematicians, I believe we'll always need more than mathematical reasoning. My belief, of course, in no way implies that we'd be acting foolishly to use mathematical reasoning as much as we can.

CHAPTER

Roundly speaking

The first time I ran into the term "mass noun" it rattled me. I thought I'd done well in three grammars: English, Latin, and French. I'd even learned a little German grammar soldiering in Europe. Yet here, in a paper by that insurance-executive turned-student-of-Amerindian-languages, Benjamin Lee Whorf, was a grammatical term I didn't recognize.

> We have two kinds of nouns denoting physical things: individual nouns, and mass nouns, e.g., 'water, milk, wood, granite, sand, flour, meat.' Individual nouns denote bodies with definite outlines: 'a tree, a stick, a man, a hill.' Mass nouns denote homogeneous continua without implied boundaries.

Mass nouns, Whorf writes, "lack plurals, in English drop articles, and in French take the partitive article *du, de la, des.*" To speak of individual instances, we must measure the mass substance out into forms, a "*cup of* water," a "*stick of* wood."

The Hopi language, Whorf says, is different. "It has a formally distinguished class of nouns. But this class contains no formal subclass of mass nouns. All nouns have an individual sense."

137

I felt all this should have been on my turf, but it wasn't. The examples were clear enough, but the point escaped me. I was hung up on a distinction, for which I had little feel, between mass nouns and individual nouns, while Whorf expounded on the significance of *not* having that distinction in Hopi, a language for which I had no feel at all.

The article that so baffled me was his justly famous "The Relation of Habitual Thought and Behavior to Language." It starts with vivid examples of fires attributed in part to what things are called. Whorf describes how careless behaviors around "limestone" and "empty gasoline drums" can reflect unspoken beliefs that "stone" won't burn and "empty" means inert. But then comes the mass-noun bit that undermined my comprehension of what followed.

I desperately wanted to understand the entire article. I must have read it half a dozen times over as many years. Gradually it got clearer. Eventually I saw my mass-noun difficulty as a gap in my mathsemantic sensitivity.

Things referred to by individual nouns can be counted: one tree, two sticks, three men, four hills. We can view such events as having a one-to-one correspondence with our object-level constructs and verbal-level designations. Precise math goes well here. Mathsemantic difficulties are minimal.

Mass nouns, however, refer to unbounded "substances," water, milk, wood, etc. We can count only after measuring: a quart of water, two pails of milk, three cords of wood, four tons of granite.

We can't count the substance itself; hence its name has no plural. Rather, we measure the substance in arbitrary units to give it form: quarts, pails, cords, tons, etc. Because all such measurements are approximations, precise math can mislead. Because units are arbitrary, mathsemantic difficulties appear.

We use math with both individual and mass nouns. But math's implied exactness doesn't apply equally. It makes sense to speak of precisely two toothpicks. "Approximately two toothpicks" sounds odd. But it sounds eminently reasonable to say "approximately two cords of wood" and overly exacting to say "precisely two cords of wood."

Physical measurements necessarily yield approximations. Although we can measure some lengths with astonishing accuracy, we can't measure any perfectly. Some inexactness, even if only one part in thousands, remains. If we try to get too fine with determining length, say of an iron rod, our object-level notion of rigid boundaries fails. We reach a level of atomic structure where the "iron rod" is in flux without exact boundaries.

I remember well my first college physics class. Our professor spent the period telling us we must put away our notions of exactness. We were going to use a lot of math, that was true, he said, but the things physics deals with (time, space, mass, force, etc.) have to be measured, and measurements are never more than approximations

The professor must have seen many students who were good at math but bad at physics. Otherwise why did he hammer on such a simple point as the difference between counting and measurement, between precision in math and imprecision in physics?

This exposure came three years before Whorf baffled me, three years in which I still hadn't put math and semantics together.

We can construct a scale of sorts representing the degree to which we can or can't have precision. In the simplest case, even as infants, we face small numbers of countable things. Two spoons. We also encounter things we can't count, but can only measure. Water flowing into a tub.

Later we start imagining countable things in quantities we can only estimate. The number of fish in the sea. Still later we imagine large-scale uncountable phenomena. The amount of lava disgorged since the last ice age.

It could be argued that expecting precise correspondence of verbal, object, and event levels creates little difficulty with small numbers of countable objects. However, such expectations run into progressively more trouble as we move up the scale.

For example, a flight attendant can count exactly how many passengers are on a plane. No trouble. But ordinarily a flight dispatcher only estimates the weight of those passengers, a figure still occasionally critical to safe operation.

Counting all the "passengers" in the United States last year creates both an uncertain count and, as we have seen, questions about what we're counting. Extending measurements of weight to objects in space creates similar uncertainties. Our ordinary notion of "weight" fails, because weight depends on earth's gravity, and we must start estimating mass.

If an idea is really important, you're likely to run into it where you least expect it. This happened to me thirty years after reading Whorf, while reading Theodore Bernstein's comments in *The Careful Writer* on the difference between "fewer" and "less." "The general rule," Bernstein says, "is to use *less* for quantity and *fewer* for number."

That really did it. A physics professor, a linguist, and now a journalist had made me feel insensitive. So I practiced. Now I can actually hear a statement like "less people showed up today" as denying our individual countability.

"Fewer" and "less" are not interchangeable. Things may cost fewer dollars, less money, but hardly less dollars, fewer money. Skim milk has fewer calories, less fat, not less calories, fewer fat.

My point is not to establish proper usage. Grammar's not my game. I'm only suggesting that hearing "less" where "fewer" works better *may* tip you to the presence of someone insensitive to the distinction between individual and mass nouns, between counting and measuring. I should know.

The gap between counting and measuring seems narrow. Yet it's a major mathsemantic divide. On one side lies the safety of exact counts; on the other, the insecurity of approximate measurements.

I can understand a math student's reluctance to leap from the safety of a self-contained logical language to the insecurity of noncountable events. Having to measure raises messy practical questions on how best to measure and even on what's being measured.

I can also sympathize with anyone stepping across the divide unawares, ingenuously expecting to find events conforming to words and mathematical practices. Perhaps some such childhood confusion of levels of abstraction, an overconfident projection of math certainties onto physical events, was the physics professor's target.

Measurement yields approximations. Expressing approximations requires rounding. Therefore, a facility with rounding might indicate competence with approximations and measurement.

Actually, we round numbers for at least three reasons: to save space, speed comprehension, and signal the degree of reliability. They often go together.

A table expressing figures "in thousands" can so state just once and then omit "000" from every number. The space saved accommodates more columns.

Accounting reports round book figures to speed comprehension. Accountants do not thereby confess ignorance, only a belief that reporting pennies confuses more than it enlightens.

Market research reports also round to save space and to speed comprehension. Unlike accountants, however, we researchers have no idea what the pennies are. We probably have little idea what the dollars are. So we make our rounding obvious to remind the reader that our figures are just estimates.

Often only a number's order of magnitude (its length before the decimal point) and its first digit or two really matter. A consultant recently estimated my new computer configuration will cost $5,742. He and I both know that means around $6,000, more or less.

Almost everybody dealing with estimates tends to show misleading detail. The U.S. Census reports population to the exact person. That doesn't make the count exact. Although we count people digitally (there are no fractional live people), when their number is great we lose the ability to count them exactly.

It's easy to be misled by presumptions of mathematical exactness. Some analysts have little sense of this. A total neophyte might take a local survey showing that 102 out of 300 households had freezers, multiply it by the town's census figure of 20,945 households, and report that 7,121.3 households (102/300 × 20,945) have freezers.

But hold on. Fractional households make no sense, so let's at least round to 7,121. Further, the household count was taken earlier and wasn't exact even then. The freezer survey has also aged, probably contains reporting errors, and wasn't ever a perfect sample of all households. So 7,121 falsely implies exactness. So does 7,120. Even

7,100 overstates accuracy somewhat, although that's probably what I'd use.

I regularly produce traffic estimates expressed to two significant figures, such as 5,700 passengers, 57,000 passengers, or 570,000 passengers. I gladly round the useless digits to get a more memorable figure that exposes my ignorance. I'd rather be judged a reasonable user of numbers than a foolish calculator.

A lot of people find rounding difficult. They hate to throw anything away. The mere fact that they've used a mathematical process convinces them that they can't adjust the result. They don't feel in charge.

Math has no final rounding formula. One good general rule is to round the final calculation to the *least* number of significant figures used in any multiplier. Thus, if you've taken 34% of 20,945 households, the rule says to round to two significant figures. Therefore, 7,121.3 gets rounded to 7,100.

Rounding involves finding the correct answer mathematically and then expressing it properly mathsemantically. For example, if you round $1.997 to the nearest penny, you should express the answer to the nearest penny ($2.00) to announce that fact. If you round $1.35 to the nearest dollar, express it to the nearest dollar ($1). In rounding to the nearest hundred dollars, you have to keep the zeros (for example, $300), but whatever you do, don't add any pennies.

As a consultant, I must guard against potentially misleading details. I also try to protect those who might otherwise get lost in the numbers and follow mathematical results into never-never-land.

In our aviation business, we sometimes survey local airport accessibility. We measure driving times as best we can to the nearest second on a watch and distances to the nearest tenth of a mile on a car's odometer. We run most routes several times and average the findings. We occasionally supplement the figures with mileages scaled from maps.

Such practical work generates a healthy respect for the pitfalls of even the simplest measurements. It's especially useful for awakening intelligent new recruits. Many businesses, I suspect, have similar initiations.

Most of our work uses published figures compiled by others. Having compiled and published my own, I can spot signs of trouble and then guess, as the saying goes, where the bodies are buried. Poorly handled rounding is one such sign.

Indeed, how well people round can tell a lot. I noted earlier that measurement yields approximations and that expressing approximations requires rounding. It would seem to follow, then, that competence with measurement would promote competence with rounding. I never met a carpenter who couldn't decide what length to read from a rule. But I've known people to stumble trying to express the height of a table, a loved one's feverish body temperature, or how long a speech ran. They seemed annoyed that what they measured didn't come out in exact units.

Competence in rounding would seem to force awareness that statements—in this case, numbers—are not the events they describe, can never say all about those events, and are selected by us. Conversely, an inability to round numbers would seem to reveal at least a lack of practice expressing measurements, perhaps a more general lack of confidence in describing events numerically, and just possibly even more serious mathsemantic problems.

Therefore, I included fourteen rounding problems in my quiz. You've already seen how one of these, rounding to zero, revealed trouble. I'd now like to tell you briefly about the others.

To make this easier to follow, I'll take the quiz problems in their order of difficulty for the one hundred ninety-six applicants, easiest problems first. (If you happen to want to test yourself, you'll have to see problems 4 through 6 in appendix A before reading further; in this case I'm going to give the answers right with the problems.)

One hundred eighty-four applicants, or fifteen out of sixteen, correctly rounded 271.9 to the nearest whole number (272). Others gave ".272," "2.72," "270," "271.5," "272.0," "280," or no answer. We gave credit for "272." with the useless decimal point, but counted "272.0" as wrong because its zero promises false accuracy.

Six fewer, or one hundred seventy-eight, rounded 63.1 to the nearest whole number (63). Apparently it's harder to discard than to add a little something.

Eighteen fewer, or one hundred sixty, correctly rounded 3.5 to the nearest whole number (4). Eight rounded down (to "3") and some argued plausibly for alternately rounding .5 up and down. But we round .5 up. So do the most popular computer spreadsheet and database programs, and so has every place I've ever worked. Nine applicants used a misleading zero ("4.0").

Still fewer, one hundred forty-four, about eight out of 11, could round "7 thousand" to the nearest whole number (7,000). The answers ranged from ".7000" to "1 million." Eight applicants answered "10,000." Fifteen didn't answer. Apparently it's hard to translate words into numerals.

One hundred thirty-three, just over two out of three, correctly rounded .0250 to the nearest two decimal places (.03). Twenty-eight didn't answer.

Nine fewer applicants, one hundred twenty-four, correctly rounded 271.986 to two decimal places (271.99). Nineteen gave wrong answers from "270" through "272.98." Apparently the longer a number is, the more confusing it gets.

Only one hundred eleven correctly rounded 63.1 to the nearest two decimal places (63.10). Fifteen answered "63." Ten answered "63.1." Adding a zero to hold another decimal place causes trouble.

Only one hundred and five, just under six out of eleven, rounded 1.1049 correctly to two decimal places (1.10). Thirty rounded to 1.11. Eight had earlier rounded a five down; more here rounded a four-followed-by-a-nine up. Discussion revealed that, instead of rounding just once, they succumbed to successive roundings, first from 1.1049 to 1.105, and then from 1.105 to 1.11. But by that logic, forty-four and a half cents would round to a dollar (.445 to .45, to .5, to 1). Twenty-nine applicants didn't answer.

Ninety-six, not quite one out of two, correctly rounded 486,440 to the nearest hundred thousand (500,000). Thirty-one rounded to the nearest thousand (486,000). Seven rounded to 487,000. Twenty-six didn't answer. My favorite answer is "100,000," which seems to suggest we'd asked for rounding to the nearest " *one* hundred thousand."

Only eighty-six applicants correctly rounded 3,751,120 to the nearest hundred thousand (3,800,000). Twenty-one rounded to the nearest thousand (3,751,000), three to the nearest ten thousand (3,750,000), and six to the nearest million (4,000,000). Thirty-four didn't answer.

Seventy-two applicants, four out of eleven, correctly rounded "3 million" to the nearest two decimal places (3,000,000.00). It's like asking for three million dollars in dollars and cents. Thirty-eight answered "3,000,000" with no decimal places. Ten merely copied the original problem, "3 million," a tactic found throughout the quiz. Two applicants answered three billion ("3,000,000,000"). Forty-eight didn't answer.

Now the worst three rounding problems.

Forty-eight applicants, just under one in four, as reported in an earlier chapter, correctly rounded .098 to the nearest whole number (0). You may recall that the most frequent answers were "1" and ".1" and that many applicants later said they didn't consider zero a number.

Forty-six applicants, only two fewer, correctly rounded .098 to the nearest hundred thousand (0). We didn't count two answers of "000,000" as correct. What kind of number is that? Only one applicant answered "1," but another answered "10," two "100," one "1,000," and ten "100,000." Twenty-four answered ".1" in five different styles, of which ".100" was the most common. Sixty-eight didn't answer.

Rounding −8,763,429,019.678 to the nearest hundred thousand (−8,763,400,000) won out as the most difficult rounding problem. Applicants gave seventy-six digitally different answers, ranging from "−9,000,500,000.000" to "9,000,000,000.000." Only thirty-eight applicants, not quite one in five, answered correctly. Fifty-two applicants gave no answer.

These results point to several conclusions, some of them quite obvious.

Long numbers give more trouble than short ones. Perhaps you know the old political joke, which I first heard attributed to Chicago's first Mayor Daley, and which goes something like this: "My opponent proposes to spend the outrageous sum of one million nine hundred sixty-eight thousand seven hundred and ninety-eight dollars and seventy-two cents. My budget will be only two million."

Rounding to whole numbers is easier than rounding to two decimals. Rounding to two decimals is easier than rounding to hundred
thousands. Negative numbers and zero give the most trouble.

Rounding numerals to numerals is easier than rounding words to
numerals.

Many applicants don't understand positional notation. They don't understand what's meant by "whole number," "two decimal places,"
"hundred thousand." "Our mathematical language," says Elisabeth
Ruedy, "isn't clear to many people."

Many applicants couldn't handle rounding even as a purely mathematical operation. That's not a good sign for competence in rounding
measurements and large-scale estimates.

It seems inescapable that an ordinary conception of number rooted
in our identification of events, objects, and words leaves us quite unprepared to deal with any but the simplest counting.

Unfortunately, most of the serious threats to our species involve
measurement, large numbers, or both.

We can understand how coffee spilled on a rug, anybody's rug, creates a stain that one should hurry to remove. That's easy. Where we
fall down is understanding the effect of discarding millions of plastic
sandwich containers every day. Our mathsemantic cognitive habits
hide the big problems made up of a multitude of little actions.

"Careful there. You're about to spill your coffee. It's not good for the
rug. That's better. Thanks. Now, tell me once again, if you don't mind,
why I should try to do anything about ocean pollution. What's that got
to do with me?"

Proposition 17: Competent measurement requires facility
in making and expressing approximations.

Exercising our ability to select whatever words we like to communicate what we think we see out there, we might deduce a corollary from
proposition 17:

Proposition 18: Perfectionists can't measure up.

Think about it. Especcially if you happen to be as determined as I once was never to make a mistake or expose your ignorance.

Some people I've met express opinions confidently in words but refuse to make numerical estimates. I suspect they feel safe that most words are hard to pin down but that numbers could expose them. So they've decided to avoid numbers. Some few, regretfully, seem to have made lifetime habits out of being vague. I never hired any; we couldn't agree on the figures.

However, I sense that most people who shun numbers would truly like to improve their numerical facility, if they could only work up the nerve. I'm suggesting they relax, try using numbers approximately. I do it many times every day.

A six-year-old, asked by her teacher to add five plus three, immediatcly answered "seven." Asked how she got that answer, she counted on her fingers, got eight as the total, and declared, "I don't care if it's within one."

Kamii, who reports this incident favorably in her *Young Children Reinvent Arithmetic*, adds, "I had never heard such a statement from a first grader."

This little girl exemplifies what I mean by being relaxed. She's already willing to estimate first, and then check. Numbers don't scare her. I wouldn't be surprised if she grows up to be a physicist.

CHAPTER

14

A significant number

The single most useful number for a marketing analyst in the United States is our total population. Not just for itself, but for the bearing it has on almost every other number.

An analyst knowing the total population can, for example, immediately estimate the number of males and females. Of course, I'm assuming the analyst recognizes two sexes approximately in balance and can divide a number in half.

Given the total population and a little more effort, an analyst could make not unreasonable guesses at our country's total households, automobiles, and television sets. Or judge quickly whether reported figures, like seven million households (far too low) or eight hundred million cars (far too high), make sense.

So my quiz asked: "What is the approximate population of the United States?"

Our one hundred ninety-six applicants gave seventy-two digitally different answers. I'll group the answers for your convenience, but not so much as before, for a reason that will become clear.

Seventy-one applicants, or four out of eleven, gave the most frequent answer, which was no answer or "I have no idea." Although the population question came near the end of the quiz, this is a true reading. We gave everyone time to finish. Only fifteen failed to answer the next question and only seven didn't answer the last question of all. Indeed, more applicants gave no answer to the population question than to any other.

John Paulos says in *Innumeracy*, "I'm always amazed and depressed when I encounter students who have no idea what the population of the United States is." Perhaps he won't mind my suggesting either (1) not to ask, or (2) not to let it get to him. We can't have a valuable math teacher and writer going around that way.

The lowest numerical answer given was two hundred thousand, which is about half the population of Staten Island. The highest answer was six hundred billion, which is more than one hundred times the population of the entire earth.

One difficulty in reporting on this question is that, while we tested, the U.S. population kept growing. When we first gave the quiz, in 1969, the "correct" answer was about two hundred million (202,677,000, *Statistical Abstract of the United States*). Fifteen years later, when we gave the last quiz reported here, the "correct" answer was about two hundred thirty-seven million (237,001,000)

Now, because I happen to have the year each quiz was taken, I could compare each answer with the figure for its own year and indulge my statistical habit by standardizing the results in various ways. But I'll not try you with that. It's an unnecessary complication.

Let's just give full credit to anyone who got anywhere near close, say more than one hundred fifty million (150,000,000) and less than three hundred million (300,000,000). Forty-eight applicants, or not quite one in four, gave answers in this range.

Those of you into human ecology may be disheartened that so few applicants knew what the U.S. population was. Presumably they knew even less that the world population was around five billion (5,000,000,000) and climbing. Please, however, also note a math-semantic problem. Our "good at numbers" applicants confused thousands, millions, and billions.

You may recall the difficulties, discussed in the previous chapter, that applicants had with rounding "7 thousand," "3 million," and to

the "nearest hundred thousand." That seemed to show they weren't
sure what thousands, millions, and hundred thousands looked like expressed in numerals.

The population answers bolster this impression.

Here are all the answers given to the population question and the
number of applicants giving each answer. Both numerals and words
appear, as in the actual answers. An asterisk (*) indicates that some answers used the other style.

Order of magnitude	Number So
Answer	Responding
No answer	71
In the hundred thousands	
200,000	1
225,000	1
In the millions	
2 million	1
2,220,000	1
3 million	2
4 million*	2
5 million	1
6 million	2
In the tens of millions	
ten million*	2
20 million	2
22 million	1
23 million	1
30 mil*	2
35 million	1
52,00,000,0 [sic]	1
70,000,000	1
80 million	1
86 million	1

In the hundred millions

102 million	1
120,000,000	1
130,000,000	1
140 million	1
150 million*	2

(begin counting as "correct")

180 million	3
182 million	1
200,000,000*	18
205 million	1
210 million*	6
220,000,000*	10
240,000,000	2
250,000,000*	6
270 million	1

(end counting as "correct")

300,000,000*	9
350 million	1
400 million*	3
500 million	3
800 million*	2

In the billions

1 billion	3
1 1/2 billion	1
2 billion*	7
2.5 billion	1
3 billion	4
4 billion	2
5,000,000,000	1
6,000,000,000	1
7 billion	1

In the tens of billions

10 billion	2

In the hundred billions

180 billion	1
200 million thousand	1
250,000,000,000	1
253 billion	1
500 billion	1
600,000 million	1
Total	196

There's much more to these answers than just whether they're right or wrong. Let's do a bit of detective work.

First, some clues. The first two are 200,000 and 225,000. They'd both be correct answers if only they had three more zeros. The next two answers, 2 million and 2,220,000, would also be correct if only they had two more zeros. So would 20 million, 22 million, and 23 million if only they had one more zero. Ditto 2 billion and 2.5 billion if only they had one fewer zero. And ditto 180 billion, 200 million thousand, 250,000,000,000, and 253 billion if only they had three fewer zeros.

Aha! These clues suggest that applicants pay more attention to initial digits than to magnitudes.

Can that be? Let's analyze the data more closely.

We can start by looking at just the answers under one hundred million. Twenty-four applicants gave such answers. Eight of them begin with a two. That's odd. Any of nine numerals, one through nine (here's another place zero's treated differently), could have started these answers. So, if all numerals were treated equally, we'd expect only one ninth, or fewer than three answers, to start with a two. But we got eight, many more than expected.

We have to skip the group in the hundred millions. That's the order of magnitude where the correct answer falls, which could account for answers in that range beginning with a two.

So, next, let's look at just the answers in the billions. Twenty-nine applicants gave such answers. Eleven begin with a two. We'd normally expect about three to start with that numeral, so we've again got many more starting with two than expected.

Combining the low and high groups, the one below one hundred million and the one from a billion up, we'd expect six answers to begin with a two, but we got nineteen. That supports our theory that applicants pay more attention to initial digits than to magnitudes.

Can we sift the evidence further? Yes. We can look at how many significant initial digits (that is, numerals before the final string of zeros) were given in the answers.

Just three answers in the combined low-high grouping begin with three significant digits. These are: 225,000, 2,220,000, and 253 billion. All three would be correct if only they had the right number of zeros.

That's odd, isn't it? These answers are so precise, yet wrong. Again, it seems to say that applicants pay more attention to the initial digits of the population count than to its order of magnitude.

Let's try it with the answers in the low-high grouping that begin with *two* significant digits: 22 million, 23 million, 35 million, 52 million, 86 million, 1.5 billion, 2.5 billion, 180 billion, and 250 billion. Five of these would be correct if only they had the right number of zeros. We'd normally expect only one of the nine to be in the right range, not five, if the starting digits were picked at random. More support for our theory.

This analysis falls short of being an absolute proof. Nevertheless, it suggests strongly that many applicants pay more attention to how figures begin than to their order of magnitude.

Unfortunately, in representing events, a number's order of magnitude is more important than its first one or two significant digits. $45 and $450 have the same first two significant digits, but represent quite different room rates. The zero here is not nothing; it makes the difference.

Now there's a mathsemantic problem for you. The so-called significant figures, the four and the five, are less significant than the "mere" place holder, the presumably insignificant zero. The "significant" matters less than the "nothing." Terminology can really mislead, can't it?

Bankers have a good approach. They talk in terms of balances of low, medium, or high, for example, four, five, six, or seven figures.

"Low five figures" puts the balance around $20,000. "Middle six figures" puts it around $500,000. A $45 room would be "middle two figures."

If you know that the U.S. population is in the hundred millions, which is an obvious approximation, you're better informed than if you know only that the number in 1990 began with a two and a five, even though that's correct. Of course, you'd be even better off to know that the figure was in the *low* hundred millions.

If you know only that our population is in the hundred millions, that's somewhere from one hundred million (100,000,000) to just under one billion (999,999,999). If that's all you know, then your best *single* guess would be in the middle, five hundred fifty million (550,000,000). You couldn't then be off by more than four hundred fifty million either way.

In the actual case, with a population of two hundred fifty million, your five hundred fifty million guess would be about twice the correct figure.

Don't despair; that's not as bad as it looks.

Because, if all you know is that the number begins with a two and a five, then you're left to guess the order of magnitude. You might guess two hundred fifty thousand (250,000) or two hundred fifty billion (250,000,000,000), as our applicants did. Either way, your answer would be off by a factor of a thousand. That is, the true population (250,000,000) would in one case be a thousand times your guess and in the other only one thousandth of it.

It's usually better to miss by a factor of two (double or half) than by a factor of a thousand.

Lakoff says that the single digits are prototype numbers. Children can personally experience the difference between two and three and four and so on. We apparently establish vivid connections with these small numbers that larger ones never dim. Even the notoriously uncommunicative nineteenth-century German mathematician, Peter G. L. Dirichlet, on the birth of his first child, managed to wire his father-in-law, "2 + 1 = 3."

So what we find easiest to remember is that the population of the United States is "two something."

But what about orders of magnitude? How do we establish a connection with the difference between ten, ten thousand, ten million, and ten billion? Based on our test results, perhaps the best answer is the humorist's, which is, "poorly."

How do we grasp, for example, the significance of seven hundred thousand lawyers, eighteen million government employees, and two hundred fifty billion dollars in annual social security benefits?

No wonder large-scale problems puzzle people.

Three justly renowned decision theorists, David E. Bell, Howard Raiffa, and Amos Tversky, jointly pointed out that people making decisions "do not distinguish adequately between large numbers. Twenty thousand dollars," the three professors said, "sounds a lot like twenty-five thousand dollars." Yes, and I would like to add that twenty million dollars apparently sounds a lot like twenty billion dollars.

Before you decide that we're hopelessly inept when dealing with numbers, please note how far we've come. Before the 1600s, Patricia Cohen writes in her math history, "there was no word for population." Few besides tax collectors cared how many people there were. No country had a regular census until Sweden began the practice in 1749. Basic arithmetic wasn't a requirement for admission even to Harvard until 1802. Quantification in the United States then became a novelty in the 1820s, something new, a fad.

By 1858 public interest in numbers was still a growing and current topic. In that year Oliver Wendell Holmes, the father, opened his *Autocrat of the Breakfast Table* on a mathematical note, culminating with this observation:

> I have an immense respect for a man of talents *plus* "the mathematics." But the calculating power alone should seem to be the least human of qualities. . . . The power of dealing with numbers is a kind of "detached lever" arrangement, which may be put into a mighty poor watch. I suppose it is about as common as the power of moving the ears voluntarily, which is a moderately rare endowment.

We've come a long way.

> **Proposition 19:** How numbers start tells us less than their
> length.

Or, more ponderously but more precisely,

> **Proposition 19A:** A quantity's initial digits generally convey
> less information than its order of magnitude does.

Space rates

Wherever you happen to be is the center of the universe.

My consulting work has taken me to most large American cities, many medium-size ones, and some smaller ones, but never to one that regarded itself as off center.

Each city has its own way of proclaiming its position. My first aviation assignment took me to Dallas during an airport feud with its western neighbor, Fort Worth. Dallas officials regarded their city as the up-and-coming, broad-based, commercial capital of Texas, "Big D." They regarded Fort Worth as a feudalistic cow town. People in Fort Worth saw it differently. "Fort Worth," they said, "is where the West begins. Dallas is where the East peters out."

I was born in Indiana, raised in Philadelphia, and then lived for extended periods in Chicago and New York before returning to the Philadelphia area. I recall Chicagoans during my stay there taking offense when the *New York Times* used the dateline "Chicago, Illinois," as if there were another. I believe the *Chicago Tribune* retaliated by adding the state to New York datelines.

A few years later I was asked by a fellow patron during intermission at Cincinnati's "Opera at the Zoo" *where* in New York I lived. "On the

Upper East Side," I happily confessed. I got back to my seat still wondering why the conversation had shifted to Plattsburgh and Saratoga.

Philadelphians like to think of their city as the "Cradle of Liberty." Away from Philadelphia I've heard it described as just a convenient halfway point for the patriots from Virginia to meet with the patriots from Massachusetts.

Although Philadelphia in 1774 had been second only to London as the world's largest English-speaking city, by 1790 it held fewer than New York's 33,000 inhabitants. Philadelphia's size continued to rival New York's for a few decades. New York became the capital of the United States briefly in 1789–90, and then Philadelphia served from 1790 to 1800 while Washington was being prepared. Opening the Erie Canal in 1825 began a new expansion of New York that left Philadelphia behind.

New Yorkers today tend to think of Philadelphians as provincial. Philadelphians refer to New York as "a nice place to visit, but I wouldn't want to live there." I don't agree. I find New York demanding to visit, but quite livable once you're used to it. Either way, New York's right there on Philadelphia's horizon.

Intercity distances matter in aviation. Therefore I put two mileage questions in the quiz. I tried to pick ones that were fair, involving places known to all Philadelphians.

The first mileage question was: "About how far is it from Philadelphia to New York?"

The one hundred ninety-six applicants gave seventy-eight digitally different answers. As usual, I'll group them for your convenience.

Eleven applicants, one in eighteen, answered in terms of time, from "1 hr 10 min by train" to "4 hours by car." Four of the eleven also included distance estimates: "60–70 miles," "110 mls," "165 miles," and "350 miles." We counted none of the eleven as correct.

Twenty-one applicants, one in nine, gave answers in plain numbers without specifying units. Of these, one was at "90," four at "100," and the remainder were higher, up to "300." We counted none of these as correct.

Fifteen applicants, one in thirteen, gave no answer. We counted none of these as correct.

The remaining one hundred forty-nine applicants gave answers in miles (variously expressed as m., mi., mls., etc.). These answers distributed as follows.

Miles	Number of Answers
18	1
37	1
- -	
70	1
75	1
80	2
85	1
88	2
90	24
90–100	1
93	1
95	1
98	1
100	24
- -	
110	6
120	12
125	3
130	3
135	1
150	17
160	2
170	1
180	3
less than 200	1
180–200	1

200	20
220	1
250	5
300	6
350	1
400	3
600	2
Total	149

Averages sometimes prove more accurate than individual judgments. Let's look at two kinds.

The mean average adds up all answers and then divides by the number of answers. We can substitute "190 miles" for "less than 200" in the list above and use midpoints for the answers giving ranges. This yields 23,142 miles divided by 149 answers, or a mean average just over 155 miles.

The median average is the one-in-the-middle answer. This answer is the seventy-fifth one as they are listed above; for it has seventy-four lower and seventy-four higher answers. The median is 120 miles.

Unfortunately, both averages are too high.

The Philadelphia–New York airline city-center-to-city-center distance is only 82 miles. The airport-to-airport distance is 83 miles. We treated any answer from 70 miles through 100 miles as correct.

Fifty-nine applicants, three in ten, gave correct answers. Two applicants gave answers that were too low. Eighty-eight were too high. The general conclusion: New York isn't as far from Philadelphia as most applicants think.

The second mileage question was: "About how far is it from Philadelphia to Los Angeles?"

The one hundred ninety-six applicants gave seventy-two digitally different answers, six fewer than for Philadelphia–New York, whereas one might expect an even greater dispersal over the longer distance. The difference was in those not answering, which almost doubled, to twenty-nine, one in seven, for the Los Angeles question.

Eight applicants, one in twenty-five, answered in terms of time, from

"5 hrs via" to "96 hours" (type of vehicle unspecified). Four of these answers also included distance estimates. My favorite is "12 hrs. 1250 mls," just over one hundred miles per hour, too fast by ground and too slow by air.

Thirty-eight applicants, almost one in five, gave answers in plain numbers without specifying units. These varied from "1,500" to "8,000." We would have counted some as correct had they specified miles. As it was, we credited none.

The remaining one hundred twenty-one applicants gave answers in miles, distributed as follows.

Miles	Number of Answers
250	1
600	1
1,100	1
- - - - - - - - - - - - - - - - - - - -	
2,000	7
2,200	1
2,300	2
2,400	1
2,500	6
2,700	1
2,800	2
2,900	1
3,000	75
- - - - - - - - - - - - - - - - - - - -	
3,100	1
3,200	1
3,400	1
3,500	5
3,600	1
3,800	1
4,000	2
3,000–5,000	1

5,000	4
6,000	3
10,000	1
21,000	1
Total	121

The actual Philadelphia–Los Angeles city center-to-center distance is 2,389 miles. The interairport distance is 2,407 miles. The single answer at 2,400 miles, then, is the only one correct to two significant figures.

Speaking of 2,400 miles, that's how far Columbus thought he'd have to sail from Spain west to the Orient. He was wrong. The experts who opposed his trip held no flat-earth theory. They'd just estimated the ten-thousand-mile distance more accurately. Although they were right, Columbus lucked out. Sometimes it pays to act on wrong information. But it's not the best policy.

We gave credit for all ninety-six answers from 2,000 through 3,000 miles. The bulk of these answers, however, seventy-five of them, were at 3,000 miles, which is six hundred miles too high. You could argue that we were too generous. Yet had we defined "correct" as, say, within three hundred miles, from 2,100 through 2,700 miles, only eleven answers would have qualified. That seems too tough. At least too tough on John Paulos, who also has claimed to be "always amazed and depressed when I encounter students who have no idea [of] the approximate distance from coast to coast."

Philadelphia to Los Angeles, of course, isn't quite from coast to coast. Still, no such distance reaches three thousand miles. New York to San Francisco is 2,574 miles. Boston to Seattle is 2,495. Jacksonville to San Diego is only 2,089. Even the Miami–Seattle diagonal is only 2,726. The Bar Harbor–San Francisco diagonal runs 2,805 miles. The even-more-diagonal Bar Harbor–San Diego distance is only 2,720.

A generalized conclusion: The United States isn't as wide as most people think.

Only three applicants gave answers below two thousand miles. Twenty-one, one in nine, gave answers above three thousand miles, much too high. None, at least, exceeded the circumference of the earth, although "21,000 miles" comes close.

The next highest answer, "10,000 miles," reminds me of a particularly graphic love lyric of Robert Burns.

> And fare thee weel, my only luve!
> And fare thee weel, a while!
> And I will come again my luve,
> Tho' it were ten thousand mile!

Burns's sentiment is earthy. The farthest lovers can separate on earth before they start getting closer again on the return side is slightly over twelve thousand miles. But, taking Scotland as the center of Burns's universe, "twelve thousand mile" would have limited the lover's location to the ocean well south of New Zealand, not a practical spot. "Eleven" wouldn't scan. So "ten" it is. Math with meaning.

When it comes to giving math meaning, however, poetry must take a back seat to physics. No field has been more completely mathematized than physics. At the same time no scientists ask themselves more than physicists what they're talking about. Indeed, one could argue that semantic and mathsemantic concerns have been the major common theme of twentieth-century physics.

"As a young student," wrote Mach in 1903 shortly after retiring as a physics professor, "I was always irritated with symbolic deductions of which the meaning was not perfectly clear and palpable." That irritation may explain his campaign to ground geometry and physical concepts in sense experience, to demonstrate the "truth that abstract concepts draw their ultimate power from sensuous sources."

"This sound and naive conception of things," said Mach, "vanished and the treatment of geometry underwent essential modifications when it became the subject of *professional* and *scholarly* contemplation," which attempted to reduce its "initial principles to a minimum, as is observable in the system of Euclid. Through this endeavor to support every notion by another, and to leave to direct knowledge the least possible scope, geometry was gradually detached from the empirical soil out of which it had sprung," creating, added Mach, a system "more fitted to produce narrow-minded and sterile pedants than fruitful, productive investigators."

In 1905 and 1916 Einstein's special and general theories of relativity marked a crisis in physical thought.

"It was a great shock to discover that classical concepts, accepted unquestioningly, were inadequate," wrote Percy W. Bridgman in 1927 in *The Logic of Modern Physics*.

> Now here it seems to me is the greatest contribution of Einstein. . . . He has essentially modified our view of what the concepts useful in physics are and should be. Hitherto many of the concepts of physics have been defined in terms of their properties. An excellent example is afforded by Newton's concept of absolute time. . . .

>> Absolute, True, and Mathematical Time, of itself, and from its own nature flows equably without regard to anything external, and by another name is called Duration.

> Now there is no assurance whatever that there exists in nature anything with properties like those assumed in the definition. . . . It is a task for experiment to discover whether concepts so defined correspond to anything in nature, and . . . we find nothing in nature with such properties.

Bridgman then describes the new way of understanding concepts. "In general, we mean by any concept nothing more than a set of operations." He calls such definitions "operational" and illustrates them in a fifteen-page discussion of the question "What do we mean by the length of an object?"

In Korzybskian terms, an operational definition is a kind of extensional definition, a recipe.

"To find the length of an object," says Bridgman, "we have to perform certain physical operations." He then describes a simple task,

measuring a house lot. We lay down a measuring rod, mark where its
end falls, move the rod along to that point, and repeat as necessary until we reach the end of the lot, which we then say has a length equal
to so many measuring rods. Specifying the procedure in full, of course,
is more complex than this.

Bridgman next takes up the added difficulties of measuring a moving object, such as a street car. What if it's going too fast to board? This
procedure will necessarily be more complex than the previous one.

What if we must deal with higher velocities, such as of a star? What
operations do we use then? Bridgman now takes a full page to describe
"briefly" Einstein's operations for measuring the length of a body in
motion. Bridgman's emphatic conclusion: "Since Einstein's operations
were different from our operations above, *his 'length' does not mean the
same as our 'length.'* "

This covers lengths in motion. Now, what about large objects?
Bridgman describes a surveyor's optical measurements and triangula-
tion. What about extremely distant objects? What about submicro-
scopic lengths? Each requires different operations.

We try, of course, to cross-calibrate our different operations in
ranges where their uses overlap, so their results will agree. But, says
Bridgman, "We must recognize in principle that in changing the operations we have really changed the concept, and that to use the same
name for these different concepts over the entire range is dictated only
by considerations of convenience, which may sometimes prove to have
been purchased at too high a price in terms of unambiguity."

So "length" is like "fruit." Just as there are apples and oranges,
there are stationary lengths and high-velocity lengths, tactile lengths,
optically triangulated lengths, and ultramicroscopic lengths.

"Advancing a new theory," wrote science-philosopher Philipp
Frank, now "involved two tasks . . . : the invention of a structural system, and the working out of operational definitions for its symbols."
Einstein, said Frank, realized that geometry could be "either a structural system with arbitrary axioms or a physical theory. In the first
case, the conclusions of geometry are certain but do not tell us any-

thing about the world of experience; in the second case, the proposi-
tions of geometry can be checked by experiment" but cease to be certain.

In Greek mathsemantics, the terms of Euclid's geometry were taken
for granted and its results considered true in both a mathematical and
a physical sense. This implied that one could reason about the world
independently of experience. Hence much mischief was done, which
Mach deplored and mathematician-author E. T. Bell characterized
vividly.

> The cowboys have a way of trussing up a steer or a pugna-
> cious bronco which fixes the brute so that it can neither
> move nor think. This is the hog-tie, and it is what Euclid did
> to geometry.

Einstein demonstrated, said Frank, "that there is no statement that
is derived by reasoning without sense observations and that at the
same time tells us something about the external world."

As already noted, Bridgman's operational definitions involve extensionalization. Let's try it on the distance between Philadelphia and
New York. Let's ask what might be meant by the question, "About
how far is it from Philadelphia to New York?"

The question could mean how far is it in crow-fly miles from the center of New York to the center of Philadelphia, according to someone's
definition of the centers. If we take the centers to be 9th and Chestnut
Street in Philadelphia and 8th Avenue and 33rd Street in New York,
then the correct answer is 82 miles. I'm quoting the city-center definitions and distance used in the 1967 Civil Aeronautics Board origin-
destination survey of airline passengers. You might want to use
different centers.

The question could just as easily mean how many road miles are
there between the same centers. Several applicants referred to their
actual trips and apparently took the question this way. It could then
mean by the shortest route, or the fastest route, or some route with
low tolls. These operations could generate a lot of different distances.

The *Rand McNally Road Atlas* figure was 92 miles in 1961, 93 miles in 1971, 106 miles in 1981, and 101 miles in 1991. Don't ask—I don't know.

The question could also be taken as referring to airport-to-airport distance, as the origin-destination surveys have done since 1968. We must then determine how to deal with a combination of two airports at the Philadelphia end and three at the New York end (one of which is in New Jersey). I'll spare you the complications. Fortunately, the reported result for crow-fly miles is 83, hardly different from the center-to-center reading. Allowing ten more miles for surface travel would get us to 93 miles.

But the question could also mean how far is it from Philadelphia to New York at their *closest*. Again, the answer varies depending on what is meant by "Philadelphia" and "New York." We'd get one answer using small areas, the cities proper, another using what the census today calls "Primary Metropolitan Statistical Areas," and still another using the largest areas, "Consolidated Metropolitan Statistical Areas." By the latter definition the distance between the Philadelphia and New York areas falls to zero. They meet both in New Jersey and along the New Jersey–Pennsylvania border.

The question might even be taken to mean how far apart any point in the Philadelphia area is from any point in the New York area. Distances could then run from Delaware into Connecticut, over two hundred miles.

Now, what difference do these differences make? Are they just academic or do they really matter? Is such mathsemantic sensibility really needed?

My answer: Yes. I'll give you just one example.

It is sometimes proposed that aircraft needing shorter runways (including helicopters, which are capable of vertical takeoff and landing, VTOL) could reduce congestion at large airports by diverting their short-haul traffic to smaller fields in the same metropolitan area. The proponents mistakenly forecast that the new airports will divert traffic without generating any significant amount of entirely new traffic.

The underlying mathsemantic error is in treating specific *points* (airports) as equivalent to *areas* of the same name. In childhood's name-equals-thing semantics, "Philadelphia" can simultaneously mean the Philadelphia International Airport (PHL) and the Philadelphia market. The two become confused, so that service *at* PHL is taken as service *at all* of Philadelphia. However, service between several new "Philadelphia" airports and, say, several new "New York" airports would serve *more* of Philadelphia and *more* of New York, thereby tapping new traffic potentials. It seems obvious, yet it gets overlooked.

Most parties to the Northeast Corridor VTOL Investigation in 1969 produced Philadelphia–New York estimates only in the tens of thousands of passengers, near the historical number. I sponsored the Philadelphia exhibits estimating annual passengers in the millions, which the judge accepted. "Actually," said the judge, "the Philadelphia parties provided convincing testimony that their estimate is the most reasonable."

If the others are right, one can reduce delays at major airports by moving local traffic to a system of short-runway airports. If my 1969 approach is right, local traffic will multiply sharply and worsen airspace congestion. Before acting, we should at least make sure what our numbers refer to. That means mathsemantic extensionalization.

Just as I found aviation analysts confusing points with areas, Orr found black-English-vernacular-speaking students confusing location with distance. Just as I found that some applicants used time-units to answer distance questions, Piaget found that in children "the temporal order is confused with the spatial order."

Apparently our ability to handle physical measurements starts off awkwardly. Piaget reports that children see velocity in terms of the act of overtaking. The evolutionary survival advantage seems obvious.

The Greeks were also tied to their immediate perceptions. They believed, for example, that moving objects not acted upon by force tend to come to rest. Perhaps most people still do. It takes a release from one's here-and-now to recognize that unseen forces, such as friction, bring ordinary moving objects to rest.

Newton's great contribution to our understanding of space and time was freeing us from our immediate perceptions. He enabled us to understand both terrestrial and celestial motion in terms of more general principles that could be quantified and tested by experiment.

Understanding those principles, however, requires a mathsemantic outlook found neither in children nor in the ancient Greeks. We need to transcend our own sense impressions, to abandon the notion of being at the center of the universe.

You might want to test your own mathsemantic development in the everyday physical area quickly through just these five questions.

1. What is length times width called?
2. What is length divided by time called?
3. What is area times height called?
4. What is speed divided by time called?
5. What is volume divided by time called?

The answers. Length times width is called area. For example, a space three feet long by four feet wide is an area of twelve square feet.

Length divided by time is called speed or velocity. For example, a length of one hundred miles divided by two hours yields a speed of fifty miles per hour. "Per" means "for each," which is another meaning of "divided by." A six-pack divided by three persons is two cans for each.

Area times height is called volume. For example, an area of twenty acres times a height of four feet, as in a small reservoir, contains a volume of eighty acre-feet.

Speed divided by time is called acceleration. The average acceleration of a Corvette jumping from a standstill to sixty miles per hour in five seconds, for example, is twelve miles per hour per second, meaning that it's increasing its speed at that rate. Speed is distance over time, as in "sixty miles *per* hour." Acceleration is distance over time over time, as in "twelve miles per hour *per* second."

Volume divided by time is called rate of flow. Thus, if we emptied our reservoir's eighty acre-feet in four hours, the water would flow out at the rate of twenty acre-feet per hour.

If you got all of these right, you're either into more advanced things or ready for them. If you got fewer than three right, cheer up. There's a whole new mathsemantic world just waiting to give you pleasure and broaden your understanding of what's going on.

If you want, go back over the five questions until you've mastered them. Then use that insight to start fathoming other physical relationships and rates as they come up (such as pressure, force, growth, depletion, why divers get the bends, how the length of the day changes, $E = mc^2$, and all that).

No one need confuse, like some flight attendants, altitude with pressure. Perhaps you've heard oxygen-mask instructions that begin, "In case of a loss of cabin altitude." That seems to mean "if the plane is dropping." However, what's meant is "in case of a loss of cabin pressure," as from a hole, which would thin the oxygen to that of the surrounding atmosphere. If you're high enough, you'll need an oxygen mask. There's a world of difference.

Proposition 20: Physicists are natural mathsemanticists.

CHAPTER

Caring for symbol systems

The transmissions reported below began at five hours, three minutes, twenty-nine and three-tenths seconds after the hour of noon on the twenty-seventh of March, 1977, at the airport on Tenerife Island in the Canaries.

> **1703:29.3—Pan Am N1736 to Control Tower—**Would you confirm that you want the clipper one seven three six to turn left at the *t h i r d* intersection?
>
> **1703:36.4—Tower to Pan Am N1736—**The third one Sir, one, two, three, third third one.
>
> **1703:39.2—Pan Am N1736 to Tower—**Very good, thank you.
>
> **1703:47.6—Tower to Pan Am N1736—**... er seven one [sic] three six report leaving the runway.
>
> **1704:58.2—Tower to taxiing planes—**... for your information, the centerline lighting is out of service.
>
> **1705:05.8—KLM 4805, acknowledging—**I copied that.
>
> **1705:07.7—Pan Am N1736, acknowledging—**Clipper one seven three six.

1705:44.8—KLM 4805 to Tower—Uh, the KLM . . . four eight zero five is now ready for takeoff . . . we're waiting for our ATC [air traffic control] clearance.

1705:53.4—Tower to KLM 4805—KLM eight seven [sic] zero five, uh, you are cleared to the Papa beacon climb to and maintain flight level nine zero . . . right turn after take-off proceed with heading zero four zero until intercepting the three two five radial from Las Palmas VOR.

1706:09.61—KLM 4805 to Tower—Ah—roger, sir, we are cleared to the Papa beacon flight level nine zero, right turn out zero four zero until intercepting the three two five. [throttles opened, starting takeoff] We are now at takeoff.

1706:18.19—Control Tower—O.K.

1706:19.39—Pan Am N1736 to Control Tower—No, uh, and we are still taxiing down the runway, the Clipper one seven three six. [overlapped by the next tower transmission, making **both** unintelligible to KLM 4805, now on its takeoff roll]

1706:20.08—Control Tower to KLM 4805—Stand by for takeoff, I will call you.

1706:25.47—Control Tower—Papa Alpha [PA] one seven three six report runway clear.

1706:29.59—Pan Am N1736—O.K., will report when we're clear.

1706:32.43—within the KLM cockpit—Is he not clear, then?

1706:34.10—What do you say?

1706:34.70—Is he not clear that Pan American?

1706:35.70—Oh, yes (emphatic).

1706:47.44—[The captain utters an exclamation—followed shortly by sounds of impact.]

Five hundred eighty-three persons died as the KLM Royal Dutch Airlines Boeing 747 on its takeoff roll collided in the fog with the Pan American Boeing 747 that had not yet exited the runway.

Among the causes reported by the Spanish Ministry of Transport and Communication was the ambiguity of KLM's, "We are now at take-off." The tower, not having issued takeoff clearance, interpreted KLM as meaning, "We are now at takeoff *position*." Pan Am apparently had enough doubt to wish to make its own whereabouts clear. The KLM crew must have felt they had communicated clearly. There's no way to ask them.

On January 25, 1990, an Avianca Boeing 707 ran out of fuel about a half hour after missing an approach at New York's John F. Kennedy International Airport and crashed on Long Island. The principal cause reported was the copilot's failure to use the magic word "emergency" in reporting that the flight was low on fuel. The captain had requested, *"Digale que estamos en emergencia."* ["Tell them we are in emergency."] Unfortunately, the copilot had advised the tower, "Once again we're running out of fuel." Pilots frequently jockey for priority, so controllers tend to discount nonemergency pleas.

On July 31, 1990, an air traffic controller declared a fuel emergency for an Avianca flight when he was unable to ascertain whether the flight had only fifteen minutes of fuel remaining. The plane was vectored from Washington airspace directly through to John F. Kennedy International Airport ahead of other traffic. After landing it was determined that the flight had enough fuel left to fly another two hours.

"Communications," said Air Line Pilots Association spokesman Thomas Kreamer, is the "most fragile link in our safety chain."

"Where on earth is Baltimore, Indiana?"
"Baltimore, Maryland?"
"No, Baltimore, Indiana."
That's the question clerks asked me while tabulating Dallas hotel registration data.
In Dallas, we'd posted over one hundred thousand hotel registrations by visitors' home towns. The idea was that finding where Dallas

visitors came from would help determine where better air service might be needed. Now back in New York we had to make sense of the data. We had to find every home town and assign it to an air service area. Yonkers belonged to New York. Pasadena belonged to Los Angeles. But where did Baltimore, Indiana, belong?

None of our maps or reference books showed Baltimore, Indiana. Deciphering handwriting, some of it so bad it seemed intentional, often caused problems like this. Perhaps it was really "Bartholomew," an Indiana County. Then it would belong to the air service area of Columbus, Indiana. Or might it be Burlington? That was near Kokomo. None of these seemed sure enough. We hated to introduce a bias by giving up, so we put the problem in the "do later" pile.

When at last we tackled the toughest problems we found we really had to solve "Baltimore, Indiana," because it had shown up many times. We couldn't just invalidate the faulty records; for that could seriously underestimate Dallas's need for service somewhere. But where? Fortunately, we'd also stumbled onto "Chevy Chase, Indiana," "South Bend, Maryland," and "Fort Wayne, Maryland." That made it easy. The handwritten abbreviations "Md." and "Ind." are often indistinguishable. Our clerks in Dallas, apparently not too good at geography, had misread the abbreviations.

Something similar, but troubling to more people, happened when newly opened Dulles International Airport was given the code "DIA." Washington was also served by National Airport, coded "DCA." Any hurried check-in clerk can write "DCA" and "DIA" so they're indistinguishable to a baggage handler. You know what happened. Washington travelers got separated from their luggage. That's why Dulles International Airport is now coded "IAD."

The Federal Aviation Administration (FAA) assigns aircraft identifications. They begin with N, continue with a set of from one to four numerals, and may end with one or two alphabetic characters. The FAA, however, will not permit the letters "I" or "O" to be used, for fear of confusion with the numerals "1" and "0." Further, the FAA will not permit the first numerals after the "N" to be zeros, for fear of "N0012" being reported as "N12." The resulting confusion could affect air traffic control, regulatory enforcement, and filing, presumably in that order of importance.

Any language, handled sloppily, can give false leads. Error could endanger me on the witness stand, so I paid attention.

Ordinary words often withstand sloppy handling. Because they're phonograms, you kan git the meenings even when therr hoaribly misspeld. One letter seldom makes a difference. The worst single-letter error may be interchanging "now" and "not." The switch reverses the meaning with emphasis. Imagine opposing something but filing a pleading, "We do now support the motion." Or favoring it but filing, "We do not support the motion."

Punctuation requires more care, especially commas.

An alumnae invitation to a Christmas celebration suggested, "Bring your children and a tree ornament, to be donated to a charity." Sorry, we love them too much.

Mathsemantics requires handling these ordinary problems with more than ordinary care. There are two main reasons. First, few mathsemantic errors are as obviously laughable and unpersuasive as a grammatical implication that we donate our children to charity. Second, math packs more meaning into less space.

Math symbols, as compact ideograms, require constant care. Omitting one digit changes 200 to 20. Misplacing a period changes 1040.1 to 10.401. Scribbling nines too fast can reduce them to fours. Superimposing the base of a "2" on a printed line can make it look just like a "7."

Do poor spellers gain from sloppy handwriting that hides whether an "i" comes before or after an "e"? Does anyone gain from a "6" written over a half-erased "5" that now could be read as either? Or a "7" written over a half-erased "8"?

I'd had my share of math afflictions. I set out to hire people "good at numbers" whose own bad math experiences might keep them from adding to mine.

I crafted the first two problems on my quiz to that end. Both came

under the general instruction, "Solve the following problems in addition." The first problem was the more difficult mathematically but
without mathsemantic complication.

```
                    1.32
                   21.06
                    1.07
                    9.83
                    5.26
                   ------
```

One hundred seventy, almost seven out of eight, of the one hundred
ninety-six applicants gave the correct answer, "38.54."

Another five applicants used the same numerals, but erred by con-
juring up a dollar sign ($38.54) or omitting the all-important decimal
point ("38 54" and "3854").

The remaining twenty-one applicants just added incorrectly. Most
answers came close; some, excruciatingly close. All except one,
"28.54," fell within one of the correct answer. That's a small error as
errors go. On the low side these were "37.54," "38.44," "38.45,"
"38.47," "38.48," "38.52," and "38.53." On the high side, "38.55,"
"38.56," "38.64," "38.74," and "39.54."

The second problem, shown below at left, is mathematically simpler
than the first problem. I'm sure of this because forty-one applicants re-
cast it in the form at right (using the ample margins we suggested
were for calculations) and forty of them got it right.

```
          2.0                         2.00
           .30                         .30
            .9                         .90
          106                       106.00
         -----                     -------
```

However, you can see that I deliberately loaded the problem with
two housekeeping complications. I expressed its figures to different

numbers of decimal places and misaligned its decimal points. The cleaned-up version on the right could appear in a math workbook. The version at the left mimics what I'd found in analysts' work-papers.

The housekeeping complications had two calamitous effects.

Calamity number one was that only one hundred and two applicants, just slightly over half, got the right answer ("109.2" or "109.20").

Calamity number two was that many answers were absurdly wrong. Just twenty-four were close ("108.12" through "109.39"). The other seventy included the quiz's first "no answer" and sixty-nine more varying from ".0345" to "1092." If this were money, the answers would range from about three-and-a-half cents to one thousand ninety-two dollars. Some other odd answers were "12.39," "3.26," "318," "4.18," and "5," all seriously flawed.

The applicants giving such answers obviously didn't follow two of Elisabeth Ruedy's strategies for problem solving. They didn't rewrite the problem. And they didn't make a quick estimate. They abandoned judgment. If they'd just guessed sensibly, wouldn't they have seen that the answer *had* to be somewhere around one hundred and eight?

The most frequent wrong answer ("3.45") was given by forty-two applicants who apparently ignored the decimal points and added the columns exactly as they lay.

$$
\begin{array}{r}
2.0 \\
.30 \\
.9 \\
\underline{106} \\
345
\end{array}
$$

Then, apparently feeling a need to stick a decimal point *somewhere*, they seem to have taken their cue from the first number, "2.0," hence "3.45."

Twelve others followed the same columnar procedure but put the decimal point *elsewhere* to produce ".0345," ".345," "34.5," and "345." So we got to see the numeral sequence "345" with five different decimal-point locations. This seemed to say that these fifty-four applicants, three in eleven of all applicants, had only the haziest notion of what a decimal point means.

A retired mathematics teacher, on seeing some of our quiz results, suggested how questions could be put to get more right answers. I'm sure he's right, but he missed the point. I wasn't going to be judged by how well our employees did on an exam. Rather, we were *all* going to be judged in the marketplace by the cost of our services to clients and by how well we survived the witness stand. Efficiency forbade my hiring workers for whom I had to line up decimal points. I needed ones who could take raw data and line up the decimal points themselves.

Now, whereas the placement of a period affects the meaning of a numerical entry, the commas separating millions from thousands from the rest in large numbers, as in 3,685,019, merely make them easier to read. This comma is optional in numbers below ten thousand, such as 5019, and is not used in years, 1066, 1776, or 2001.

Although I've long felt it wise to use the comma in large numbers, we imposed no penalty for its absence in grading the quiz. Applicants not using commas penalized themselves.

One quiz problem called for multiplying 1246 by 26. Sixty applicants, about three in ten, gave answers with a comma, for example, "32,396." The other one hundred thirty-six applicants used no comma, for example, "32396." Ten of the sixty answers with commas, one in six, were wrong. Twenty-seven of the one hundred thirty-six answers without commas, one in five, were wrong. That's a difference, but perhaps not enough to trust.

It would help to know how the applicants who used, or didn't use, commas made out on other questions. To make the comparison more compelling, let's take only applicants who got the multiplication answer right, either with or without a comma. That equalizes the ability of the two groups, at least on multiplication.

The results are revealing.

The comma-users did no better than the nonusers on such matters as addition with decimals already aligned (the quiz's first problem) or on such unrelated matters as multiplying by a negative number or deciding when to mail a letter.

However, the comma-users surpassed the nonusers on adding with decimals misaligned (the quiz's second problem), multiplying decimals times decimals, rounding ".098" to the nearest whole number (zero), rounding "7 thousand" to the nearest whole number, and rounding "3 million" to the nearest two decimal places.

The comma habit apparently helps clarify decimals, zero, thousands, millions, and perhaps the whole operation of decimal notation. Apparently it's also not the habit of the majority. Nevertheless, I'm going to stick with it.

A map's use depends on its clarity.

For example, unless otherwise instructed, persons posting data to tables create considerable ambiguity in their use of blanks, dashes, NAs, and zeros. Say the table is reporting passengers exchanged in a twenty-station matrix; that is, in a table listing twenty stations vertically paired with the same twenty stations listed horizontally. You run into a blank entry space. Which of the following does it mean?

1. I didn't get this far.
2. I skipped this one.
3. There were no passengers.

You can't be sure. Therefore, to be safe you might have to redo the work.

Next you come across a dash. What does it mean?

1. There were no passengers.
2. I couldn't find any data.
3. No entry belongs here.

Again, you can't be sure.

Then you come across "NA." What does it mean?

1. Not available (as in data for that period have not yet been published).
2. Not applicable (as for the box in the matrix which is Chicago both vertically and horizontally).

You have to take another look to be sure.
Then you come across a zero. What does it mean?

1. I didn't find the market pair.
2. I looked everywhere the market pair could be in the source document and there were no passengers reported.

For example, if the market pair is Chicago-Islip, an unsuspecting worker will look up Islip under Chicago, find no listing, and post a zero, meaning no more than what's stated in item #1 immediately above. A more cautious worker, for whom a zero would have meaning #2, will try looking up Chicago under Islip, only to find no Islip at all, and ask about it. Being told that Islip is the old name for what is now usually called Long Island MacArthur, the cautious worker then looks up Long Island MacArthur under Chicago and finds plenty of traffic.

I've seen these ambiguities recur often enough in my own practice and elsewhere to estimate they occur millions of times per day around the world. My experience indicates they occasionally cause serious misinterpretations. Yet I've never seen them addressed anywhere except in our own office procedures.

Treating language incautiously may be a carryover of the child's confusion of language with events. Language requires some kinds of care that events don't demand. What do events have to do with decimal points, commas, and such? Nothing. Only language does.

But there is a link. If we disregard the housekeeping needs of the languages that make up our maps, we risk paying for it with disturbing events.

Proposition 21: Sloppy mathsemantic maps create real-world dangers.

CHAPTER 17

Sorry to be late

A frequent mathsemantic casualty is the notion of maximizing. Consider the following problem from our recruitment quiz.

> To have the best chance of a letter's being received in San Francisco by July 15, you would mail it on ( ) July 12, ( ) July 11, ( ) July 10, ( ) July 9, or ( ) July 8.

This was the only multiple-choice question given the one hundred ninety-six clerical and secretarial applicants. We provided space for five choices. We got eight.

Seven applicants, one in twenty-eight, didn't answer. Two applicants combined choices. One picked July 10 or 12, the other straddled July 10–11. I'll omit these equivocations.

The largest single group of applicants, eighty-eight, four out of nine, correctly chose the earliest date, July 8. Their choice gives *the best chance* from those provided *of the letter's being received in San Francisco by July 15.*

Ninety-nine applicants, more than half, chose specific dates reducing the chance of receipt by July 15.

Date	Answers
July 9	9
July 10	37
July 11	19
July 12	34
Total	99

Indeed, as this table shows, thirty-four applicants, more than one in six, would mail at the last opportunity, another one in ten would take the next to the last.

What possible reason could there be for not mailing at the earliest date? Especially here, where it costs nothing? Failure to use costless advantages must surely be the purest form of folly.

You may have noticed the jagged, up-and-down, pattern of answers. It becomes even more obvious when we include the number who picked July 8, as follows: 88, 9, 37, 19, 34.

One of the crazy-marvelous things about numbers is how they reproduce patterns you've seen before. As Yogi Berra may have said, "It's déjà vu all over again." This reprise is so far afield I have to tell you about it.

The first time I recall seeing this pattern was in 1956 when I had forecast, for what was then called the Socony Mobil Oil Company, how motorists' gasoline-grade purchases might distribute if the company changed from offering two grades of gasoline to offering five grades. At the time, Sunoco had a five-grade experiment going, and Esso had just gone to three grades. Many millions of dollars of refinery and service-station construction costs rode on Mobil's decision.

I'd developed the theory that there were three kinds of buyers. First there were "price" buyers, who bought the cheapest grade of gas even if their cars had high-compression engines. Presumably these cars sometimes lost power and knocked going up hills, but the motorists didn't care.

Second, there were "requirement" buyers, who bought the grade recommended by their car's manufacturer.

Third, there were "quality" buyers, who bought premium even if their cars had low-compression engines unable to benefit from the higher octane in premium gasoline. Presumably these motorists got satisfaction from feeling they were buying the best.

When I estimated how these two-grade buying habits would split in a five-grade market, an interesting thing happened. The lowest, middle, and highest grades got most of the business. The other two grades, the second and the fourth, got less. Lakoff might say those grades fit fewer people's idealized cognitive model of gasoline.

The jagged distribution apparently resulted from the mix of three different gasoline-buying criteria.

So, maybe three different criteria affected our mailing-date question. Perhaps one group, "early birds," maximized the chances of timely delivery by mailing as early as possible. A second group, "punctual mailers," tried to get the letter there on exactly the date given. A third group, "procrastinators," waited until the last minute.

An interesting theory. Unfortunately, we can't prove it from our quiz. There's not enough information.

What the quiz and subsequent discussion did disclose is that most incorrect answers came from applicants' trying to get the letter to San Francisco *on* July 15 rather than giving it the best chance of being received *by* July 15.

"Why," I'd ask, "would you mail the letter on July 12?"

"Because it takes only three days to get a letter to San Francisco," applicants replied.

"But you're taking a chance, aren't you, that this time it may take longer?"

"Not really," they said, "three days is enough time. It might even get there faster."

"But," I'd suggest, "if you can mail the letter whenever you like, and you want the *best* chance of its being received *by* July 15, why not mail it as early as you can?"

"I just don't think it takes more than three days. My friends in California, in Los Angeles, get my letters in three days."

"Well," I'd purr in my most ingratiating manner, "maybe you're trying to get the letter there *on* July 15, rather than *by* July 15." I figured applicants would eventually let me in on what they were really thinking.

The answer to my *on* versus *by* suggestion most often came back as a question, "What's the difference?"

And that stumped me. I'm really not too good with people who can't distinguish between taking action *on* a date and *by* a date. I tried to explain that "by" meant "no later than," but with little success. Oh, sure, some applicants went along with me, but then some applicants might follow any suggestion.

And some applicants wanted to argue about it. That's also understandable.

But I don't recall a single "Aha," a single instance when an applicant brightened with the realization that *maximizing the chances of delivery by a given date* entails mailing as early as possible.

Applicants refused to maximize even after it was explained. They kept wanting to convert the mailing problem into the question, "How long does it take to get a letter from Philadelphia to San Francisco?" They seemed to miss, somehow, the point of view inherent in maximization.

This finally led me to question my assumption that those answering "July 8" were really trying to maximize delivery *by* July 15. I checked the answers. Sure enough, two applicants picking July 8 had added remarks showing they intended delivery *on* July 15. One justified her July 8 choice with "considering today's mail service" and the other with "gauging by my experience with the mails, it usually takes 5 days just to get to Boston."

Therefore, *fewer* than four out of nine were maximizers, although I can't say how many fewer.

This mailing-date question lies in no-man's-land. It isn't pure math to those who say pure math mustn't refer to anything but its own "mathematical objects." And it isn't applied math to those who say that applied math's answers may be tested by experiment.

I see the question as mainly mathematical, but open to influence by an applicant's decisional outlook. Applicants who regarded it simply as a math question apparently did the best.

I included the question in our quiz because I'd often observed people having difficulties maximizing. Many decision analysts have noticed the same thing. There's even a term, "satisficing," introduced by Herbert Simon, to cover satisfactory and sufficient, but nonoptimal, choosing.

We all deal repeatedly with deadlines. Each applicant must have faced thousands of them regarding bedtime, getting up, mealtimes, homework assignments, family curfews, appointments, closing times, applications, telephone calls, even paying taxes on time.

Presumably, by the time I saw applicants their decision-making habits regarding time commitments were well formed. I remember one otherwise excellent member of our analytical staff who blamed her tardiness on an irresistible urge to thwart all calls for punctuality. "I can't help it," she said. "When I'm told a project must be on time, it reminds me of my parents making me get up before dawn to do farm chores."

Among our workers, maximization failure usually showed up as a lack of initiative. Our favorite assignments to get them into the habit of taking effective action were inventory control and procedural-filings-receipt control.

Inventory control involved managing our resupply system. It worked beautifully so long as its manager complied with the written procedures about monitoring receipts, following up with suppliers, and reporting delays promptly to older hands who knew how to get action. The assignment unmasked managers who feared to face facts or be assertive. Because we started with good people willing to learn, most soon overcame those fears enough to become good inventory-control managers.

Procedural-filings-receipt control was tougher. It involved insuring that we got every filing of every party in every aviation case involving any of our clients in time to take action. Its manager had to ask the Civil Aeronautics Board Docket Section who had filed and then make sure we got *all* the filings, most of which arrived without special effort.

Timely receipt mattered, for we had to review each filing, decide whether a response was needed, get a go-ahead from our client, prepare the response, send it to our client for final approval, and then either file it ourselves or forward it to our client in time for routine processing.

The parties in these cases, including our clients, were spread across the country. There were no fax machines. Some filings were thousands of pages long. Failure to meet a procedural deadline meant a disallowed response.

In thirty years we never missed a procedural date and rarely put our clients under tight time pressure. Our system paid off.

So, until a procedural-filings-receipt control manager had learned the ropes, one of our seniors kept close tabs but used a routine that allowed the manager to learn. A typical case went like this:

Monday afternoon, February 3: "Jim, I see we're still missing exhibits from three parties in the XYZ case. They were to have been filed Friday, and we must have them by next Monday."

"Right. I haven't called anyone yet, because they might just have mailed them on Friday."

Tuesday afternoon, February 4: "Jim, I see we're still missing two of the direct exhibits in the XYZ case."

"Yes. We got one today. The others will probably come in tomorrow."

Wednesday afternoon, February 5: "Jim, I see we're still missing one of the direct exhibits in the XYZ case."

"Yes, I'd better call them."

Early Thursday afternoon, February 6: "Jim, I see we're still missing those direct exhibits in the XYZ case."

"Yes, I called them, and they said they'd mail us another copy."

"Oh, does that mean they said they'd mailed one earlier?"

"Well, I didn't actually ask them that. But we should have it on time."

"Jim, when did you call them?"

"They're in Phoenix, as you know, so actually it wasn't until about noon today, after their office opened."

"When did they say they'd put the copy in the mail?"

"Well, they didn't say. I'm sure it will be today."

"Jim, can you guarantee that we'll have a copy by Monday at the latest?"

"It shouldn't take longer than that to get mail from Phoenix."

"Look, Jim, sometimes it does. I want to be sure you get my point. We've got an obligation to protect our client. We're depending on you. The missing exhibits might require extensive rebuttals. Monday is our *deadline*. Procedurally, that's already a week late. Did you check the mailing address they're using?"

"Maybe I'd better call them back and check on these things."

"Jim, look at it this way. Even if they have the correct address, even if they promise to mail today, we still can't be sure we'll receive the exhibits by Monday. They may find they're out of copies. Their mailroom may louse things up. The post office may not come through. It may snow. Even if we get the exhibits, pages may be missing. Can you guarantee we're protecting our client properly?"

"Ed, when you put it that way, you really make it sound impossible. Frankly, I can't give you that kind of guarantee. Nobody could. You have to trust people. What on earth do you expect me to do?"

Given this invitation, I'd buzz my secretary and ask her to make a reservation for Jim to Phoenix for that evening, returning the next day. "There," I'd say to Jim with a smile. "Now you can be at their office to get a copy, check it out, call to reassure us, and return in plenty of time."

And then I found myself looking into a startled face gradually going flush. My initiative seemed to be spoiling somebody's day.

"Stay calm," I'd say. "I don't want to spend money or inconvenience you any more than you do. It's just that we're obligated to protect our client fully, even if we end up paying for it because of our inefficiency. If you can figure out a less costly way to insure our having the exhibits by Monday, just let me know. Right now, however, you'll have to excuse me. I've other things I must do."

It never failed. Jim would storm out and then start thinking. To beat the challenge, he'd put his ingenuity to work rapidly calling up different space-time-people-energy-cost configurations. What was the *best*, *sure* way to get the exhibits?

In about an hour Jim would return triumphant and make me hear the whole story, blow-by-blow.

First, he'd found someone much nearer than Phoenix, usually in New York, who had exhibits Jim could go copy on Friday or Monday, if necessary. Next, he'd gotten the Phoenix people to send their exhibits immediately by special delivery. Then he'd gotten their legal representatives in Washington, who hoped we wouldn't mention any procedural delay to the judge, to send us their copy by overnight express, which they could replenish from Phoenix. Along the way he'd discovered that the missing exhibits were only six pages long. So he'd asked a friendly party to the case what was in the exhibits and determined there wasn't anything requiring rebuttal. That friend was also going to mail a copy. Three other people he'd called as insurance still hadn't called back.

"Well," Jim would then put it to me, "is this good enough? I don't have the exhibits yet, but if they don't arrive tomorrow or Saturday, I'll get them myself in New York on Monday."

In reply, I'd buzz my secretary. "You can cancel the reservations for Jim to and from Phoenix."

"Oh," I'd be told softly, "when I saw what he was doing, I made sure there were seats available, but I didn't actually make the reservations."

Proposition 22: Maximizing takes mathsemantic initiative.

CHAPTER

Time

D riving to the airport one evening in 1976, I saw an oncoming car suddenly turn left directly across my path. Given another second, I'd have missed it. As it was, I had no chance. I watched the inevitable collision develop as if in slow motion until my car plowed into the side of the other. Both cars had to be towed away. Fortunately, no one was hurt.

Most misfortunes could be avoided if only we had more time. Handling time well might be the most useful mathsemantic skill we can develop.

Some people say that knowing how to handle time is a talent you're either born with or you're not. But that's absurd. Children don't understand time. Indeed, they have more trouble with time-semantics than with either math or space.

Take age, for example. Children hear a lot about it, which adults assume they understand. Yet Piaget's studies show children don't understand age. Asked how old she was when she was born, a four-year-old said she "no longer remembered . . . it was much too long ago," and a five-and-a-half-year-old said, "I can't remember. . . . Oh yes, I was two months old!" Early trouble with zero, perhaps.

Further, children don't relate age to birth order.

> "Is your father older or younger than you?" *Older.* "Was he born before or after you?" *I don't know.* "Who came first, you or him?" *Me.*

Children think aging stops when you're grown.

> "Are you going to stay the same age all the time or will you grow older?" *I shall grow older.* "And your father?" *He'll remain the same age.* "And will your mother grow older?" *No.* "Why?" *Because she is old already.*

Just as children confuse the amount of milk with its height in a glass, and the quantity of objects with their extent in a row, so children confuse age with size.

> "Who is the older of you two [sisters]?" *Me.* "Why?" *Because I'm the bigger one.*

Age-equals-height semantics makes bonsai and really short oldsters fascinating. It also destabilizes differences in age.

> "When he was small, how many years older was he than you?" *Two years.* "And now?" *Four years.*

And one can catch up, even go ahead.

> "Are you the same age [as your sister]?" *No. . . .* "Who was born first?" *She was.* "Will you be the same age as her one day . . . ?" *Soon I shall be bigger. . . . Then I shall be older.*

Children don't understand time as a general reference system composed of a never-ending series of equal intervals. A child does not automatically sense time as continuous. A child can witness the successive lowering of colored liquid in a calibrated flask as it flows out the bottom, yet be unable to put clearly progressive illustrations of the event in correct sequence.

To grasp continuous time, the child must get free of the present, become less egocentric, be able to imagine event sequences both forward and backward. This ability grows gradually.

Adults may accept clocks and sandglasses as measuring out a general, equal-interval system, but children don't. For them, time is local and variable. Ask a child to rap slowly on a table while the sand in a glass completely runs out, then to do it again, but quickly.

"How does the sand go when you rap slowly?" *It goes less quickly.* "And if you rap quickly?" *It goes more quickly.* "But does the sand take the same time to run out?" *Sometimes it takes long and sometimes it takes less long.*

Just as children confuse age with height, so they confuse duration with distance. Show a five-year-old two toys moving down parallel tracks, starting and stopping together, but let one go faster and farther.

"Did they go on for the same amount of time?" *No.* "Which one went on longer?" [The child points.] "Why?" *Because it went further.*

Even for a child running with an experimenter, starting and stopping simultaneously, but ending five feet behind, distance overrules experience. The child will say that the experimenter ran for a longer time because the experimenter ran farther.

Distance seems to be an example of the identification by young children of duration with the amount of work accomplished. Thus, a six-year-old says that going fast "took longer," because it accomplished more.

[To a child who's drawn lines quickly for twenty seconds] "Was that a long time or a short time?" *A short time. . . .*

"And now [slowly, but still twenty seconds]?" *It took longer before.* "Why?" *Because there are [now] fewer strokes.*

Older children, however, may invert the relationship and identify duration directly with the quickness rather than with the output. An eight-year-old watches two men run, then agrees that "they started and stopped at the same time, but nevertheless contends that 'the quicker of the two spent less time because he was running faster.' "

By such findings, Piaget explodes the idea that our understanding of time is innate, instinctive, or intuitive. He demonstrates how we slowly construct our sense of time as we gain other understandings, including ones of math and space.

It makes sense that children have more difficulty with time than with either math or space. True, in math—contrary to usual childhood semantics—one and two aren't names that inhere in objects, but markers in a system of relationships. But at least a child can count objects as an entry into the semantics of the math relational system. After counting—or pecking, perhaps—sufficiently often, the child can grasp numerical relationships.

Then, too, demonstrations with numbers in math and with things in space have the advantage that they can be repeated in reverse. The child can learn, for example, that adding three to five yields eight, and then that taking the three away returns us to five. The child can learn that physical amounts survive changes in shape by pouring milk from one container to another and then back again,

But time flows relentlessly forward. How does the child come to accept, except by an act of faith, that the next five minutes will last as long as the previous five minutes? To a child, it makes sense to believe that time takes longer when one goes faster and farther. To a child, it makes sense to believe that each clock keeps its own time, a time that fits its immediate purpose. As Piaget put it, "The child [of eight to nine-and-a-half] is incapable of grasping the equality of time of two different clocks, even though he admits it verbally by affirming the simultaneity of the starting and stopping points."

Piaget writes about young children. He quotes no child ten or older in *The Child's Conception of Time*, the source for the quotes above, except as members of control groups. However, we would err to conclude that older children necessarily develop adult time-semantics. Half of the ten- to twelve-year-olds and a third of the ten- to thirteen-year-olds, says Piaget, responded like the younger children.

Thus, while Piaget describes how children gradually acquire adult time-semantics, he leaves open such questions as how many children succeed or how much facility they develop.

Eleanor Orr's algebra students who spoke BEV (black English vernacular) were fifteen- to seventeen-year-olds, yet their problems with time resemble those of Piaget's younger children. Take, for example, Orr's surprise at a student's explaining that how far something goes "depends only on how long it keeps moving," regardless of the speed.

> At first I found it difficult to think that a student would sustain this idea of time and distance to the point where he or she would actually think in terms of time units of distance. But I later realized that this is exactly what some students do.

Piaget says "Duration is identified with distance" by his five-year-olds. Orr says her seventeen-year-olds "think in terms of time units of distance." That's close enough for me.

Orr says her students confuse location at a time, time duration, distance covered in a time, and speed. Piaget's studies find similar confusions in younger children. Orr's findings, therefore, raise questions on the interplay of age, experience, and the mathsemantics of time, not just of algebra and BEV.

Perhaps Orr's BEV-speaking children lacked opportunities to understand time. Dolores Norton, of the University of Chicago's School of Social Service Administration, who investigated the home life of lower-income inner-city children, for example, found that most of their mothers "made very few references to time structure." They seldom referred to time sequences ("First put on your socks and then your shoes") or daily routines such as mealtimes and bedtimes. Norton discovered that the fewer such references, the worse the children did on a time test.

And perhaps the structure and prepositional forms of BEV create difficulty, as Orr suggests. The Whorf-Sapir-Korzybski hypothesis holds that behavior can be influenced by a language's structure. However, this has proven so difficult to establish that the hypothesis has become unfashionable. How does one ethically create and maintain experimental and control groups, say, that vary *only* in whether they speak BEV or standard English?

Adults also have trouble with time-semantics. Answers to our recruitment quiz, as already reported, showed most applicants couldn't divide one time figure by another, or multiply travelers by hours, or maximize a letter's delivery chances by selecting the earliest mailing date.

Some of our applicants, like Piaget's children and Orr's algebra students, confused duration with distance. They answered the "how far" questions in hours. I'm not making fun of them. Patricia Cohen notes that until a Frenchman developed a distance-measuring wheel in 1567, distance was typically reported in travel time. Even today, astronomers use what sounds like a time label, a "light-year," as a length unit, the distance just under six trillion miles that light travels in a year.

I'd like to know more about how time problems relate to other mathsemantic problems. For example, our quiz shows that applicants who failed to maximize on the letter-mailing question did almost as well as other applicants on algebra, rounding "7 thousand," and adding apples and oranges. They did *far* worse, however, on rounding to zero, estimating the U.S. population, and multiplying travelers by hours. I wish I knew how this worked. The interactions might prove interesting.

For example, I think I see how ordinary large-number difficulties can combine with ordinary childhood purposeful semantics to spoil people's grasp of the power of deep time.

Let me set this up with a parental brag. Children usually have difficulty handling large numbers and long time periods. You can appreciate, then, how my wife and I beamed when our younger daughter at age ten broke up a family conversation about Philadelphia's new longest-ever heat wave with, "We have to realize they've only been keeping records for a hundred years." Her reminder was numerically correct, friendly, and scientifically impersonal.

Contrast her remark with the common "Why us?" attitude about weather or, better yet, with the purposive semantics of Piaget's child who believed Lake Geneva was dug by workers who then filled it with water for the people.

The further back in time one must think, the more likely it becomes that purposive explanations will overwhelm explanations dependent on large numbers. People not feeling the difference between a thousand and a million could hardly be expected to grasp the differences in time scale between recorded human history (a few thousand years), human presence on earth (a few million years), and the earth's existence (a few billion years).

Thus, their chances of understanding evolution dwindle.

Stephen Jay Gould, in *Time's Arrow, Time's Cycle*, quotes Mark Twain juxtaposing large numbers and purposive semantics that twit us about our evolutionary ignorance. Twain's figures are now sadly out-of-date, but not his point.

> Man has been here 32,000 years. That it took a hundred mil-
> lion years to prepare the world for him is proof that that is
> what it was done for. I suppose it is, I dunno. If the Eiffel
> tower were now representing the world's age, the skin of
> paint on the pinnacle-knob at its summit would represent
> man's share of that age; and anybody would perceive that
> that skin was what the tower was built for. I reckon they
> would. I dunno.

Advertisers, journalists, and politicians, who presumably all know better, often omit references to time, and thereby either present re-curring, cyclical events improperly as single-instance phenomena or obscure their sense entirely.

Motorola, for example, advertises its cellular system with the head-line, "Metrophone^{SM} service, just $9.95." I presume they know that's a better sounding gift idea than the one buried in the small print, "$9.95 *a month.*"

Senator Wendell H. Ford testified before the Senate's Hearing on Airline Competition, "It is critical that we have an understanding of what policy-makers at these agencies [Departments of Transportation and Justice] are thinking when they exercise authority that will affect over 450 million passengers." This is our old friend, "passengers." The Senator couldn't have meant number of people, because our population's not that big. He couldn't have meant number of air trips, for that's only a one-year supply. Or did the Senator intend to say that policy effects dissipate after one year?

Alexander Cockburn reported in the *Philadelphia Inquirer* of August 7, 1990, that "Kuwait's overproduction of oil, in breach of OPEC guidelines, was—according to Saddam Hussein—costing his country $1 billion for every dollar that Kuwait's overproduction knocked off the price of oil." The article gives no clue what this means. I once worked for an oil company, but I still don't know what the statement means. It probably has something to do with *annual* revenues, but whether it means Iraq receives $1 billion less revenue from oil per year for each one dollar drop in price per barrel, or something else entirely, I can't tell without investing more time than it's worth.

Aviation Daily of October 5, 1989, reported that "Alaskan airports say they lost a total of 23 flights during 1988 as a result of the Soviet Union's opening up Transsiberian routings to commercial airlines." Question: Is that twenty-three flights *for the year, per day, per week,* or what? As an aviation analyst, maybe I should know, but I don't. There's no convention that flight counts mean, for example, "per day." If I'm unsure, what would a less mathsemantically sophisticated reader make of it?

Young adults and some older ones can't even handle their own inter-nal time cycle.

Human beings deprived of clocks, sunlight, and all other indications of time, says sleep-psychologist Richard Coleman, would shift to a twenty-five-hour day. Allowing such shifts for just a couple of days each weekend creates trouble for millions of people.

> Since our natural day length gravitates to 25 hours, it is much easier to stay up later than to go to bed earlier. In general, our 25-hour clock can be reset about 2 hours each day, allowing humans to live comfortably on a 23- to 27-hour day.
>
> Over the weekends, in fact, many Americans allow their sleep-wake cycle to follow the natural drift. . . . On Friday, many adults stay up a little later, going to sleep at midnight and sleeping in until 8:00 A.M. On Saturday night they may stay up until 2:00 A.M., and on Sunday sleep in until 10:00 A.M. . . .
>
> The problem with the pattern of weekend free-running comes in facing the next work week. On Sunday night, many weekend partygoers try to get back to their usual sleep schedule of 11:00 P.M. to 7:00 A.M. In trying to move back their sleep-wake rhythms by 3 hours, they try to live on a 21-hour day. The biological clock cannot easily adjust from its programmed pattern of 25 hours. As a result, Sunday night finds many day workers in bed at conventional bedtimes but unable to fall asleep.

Does that sound like anyone you know?

Adults sometimes forget that convention replaces science for counting days. At work, the prototypical day is the work day. If you're working the prototypical five-day week, and it's Thursday, someone promising you something in "three days" probably means next Tuesday rather than Sunday. If it's a promise you gave the boss, however, it might mean Sunday.

If a regulation is expressed in days, such as, "Answers to motions must be filed within ten days," it pays to ask what a day is operationally. It's perilous to assume you know. Even if the rule counts calendar (that is, "all") days as days, you're still not in the clear. What if the answer comes due on Sunday? The old Civil Aeronautics Board's rule was that due dates were the last working day within the time allowed. Thus, ten days to a Sunday really meant eight days to a Friday.

Adults also have difficulty with time as a dimension for planning their lives. Jean-Louis Servan-Schreiber, author of *The Art of Time*, says, "Mastering time is mastering one's self." Elliot Jaques, who coined the term "mid-life crisis," says everybody has a time horizon, which for children can be measured in minutes or hours, and in adults will typically be one of these: one day, three months, one year, two years, five years, ten years, or twenty years.

If we must master time to master ourselves, then it would seem we should pay more attention to our life spans as time horizons. In spite of uncertainties, much can be said. For example, by the time we're old enough to function as adults who realize that human life lasts on average only about twenty-five thousand days, eight thousand of them have already gone by and only about seventeen thousand are left.

As I write this, my own remaining life expectancy, according to the *Statistical Abstract of the United States*, is about five thousand days. Yes, I know both that I may not last until tomorrow and that I may still be around in a more worn condition ten thousand days from now. Nevertheless, "five thousand days" gives me a general frame for planning my activity.

Most people, I suspect, don't talk about lifetimes because it brings up death. I've known people who wouldn't even touch an obituary page, let alone talk about their own death. Most people can stomach reading about others' dying, some avidly watch killings on television. Yet few, it seems to me, want to make informed decisions about their own life spans. Maybe I'm wrong. It's hard to get people to talk about it.

Economists study time, but sometimes they don't follow their own advice. Economics professor Richard H. Thaler tells the following story on himself.

> Last New Year's Day, after a long evening of rooting the right team to victory in the Orange Bowl, I was lucky enough to win $300 in a college football betting pool. I then turned to the im-

portant matter of splurging the proceeds wisely. Would a case of champagne be better than dinner and a play in New York? At this point my son Greg came in and congratulated me. He said, "Gee Dad, you should be pretty happy. With that win you can increase your lifetime consumption by $20 a year!"

"Greg, it seems," adds Thaler, "had studied the life-cycle theory of savings."

Professor Edward Banfield has argued that criminals look at time much as Professor Thaler did his football winnings. Criminals want their gratification now. Thaler's son would invest the money and enjoy the interest. According to economic theory, this shift between current and future consumption is determined by one's personal time-discount rate.

A television minister argued for "green power." Don't just stew about black power and white power, he said. Concentrate on green power. Put some money away, no matter how little, to begin earning interest. "Stop working for money," he urged, "and start making money work for you." Great stuff. The life-cycle theory of savings stripped of its mathematical difficulties.

I prefer to be surrounded by people with time-semantic sophistication, people able to follow the minister's advice. They're more self-reliant, more autonomous, safer to be around. No, of course I don't believe in saving everything; I believe in having some fun.

Because time enters so many calculations, I find time-semantic savvy a definite plus. Part of that savvy is grasping how we mix mathsemantic fact and convention.

Let's try a "what if" approach.

What if it had just so happened that the moon circled the earth in thirty days and the earth circled the sun in three-hundred-sixty days? What consequences would that have had?

Well, first of all, if the moon circled the earth in *exactly* thirty days and the earth the sun in *exactly* three-hundred-sixty days, *no one* would have missed the fit.

Then months would be exactly thirty days, right? None of this sometimes twenty-eight, twenty-nine, thirty, or thirty-one days. Goodbye, one nursery rhyme.

And twelve thirty-day months would make exactly one year, right? No leap years. No gradually creeping calendars for popes to straighten out. None of this crazy business of George Washington's birthday being February 11, 1731 (old style) and February 22, 1732 (new style). No differences like Jewish years varying from three hundred fifty-three to three hundred eighty-five days, while Muslim years run either three hundred fifty-four or three hundred fifty-five days, thus making one age faster in Muslim years than in Christian or Jewish ones.

Too much regularity for your taste? You want cultural and religious differences? Okay, what would have happened to weeks?

Why do we have weeks, anyway? Koehler's birds tell us: Thirty falls outside our immediate-perception range. We need a grouping less than a month to bypass this weakness.

Some cultures would have split the months into three ten-day weeks. That's sure; for both ancient Egypt and revolutionary France did just that. However, ten also falls outside our immediate-perception range, which seldom goes as high as eight. That may explain why the ten-day model failed in the actual world. But I suspect a few highly logical cultures would succeed in preserving ten-day weeks in our "what if" world.

What weeks would the rest of us have? Would we use seven-day groupings? If we did, we'd end up with something over 4.285714 weeks in a month and something over 51.428571 weeks in a year. Given the way we and our ancestors disliked uneven ratios, I don't think seven would have had a chance.

Next up comes six. Now we're getting somewhere. Five six-day weeks fit exactly into a thirty-day month. So do six five-day weeks. So, take your choice.

And that's it, weeks would be either six or five days. Four is out, it doesn't divide thirty evenly. *Ten* three-day weeks and *fifteen* two-day weeks once again give us numbers outside our immediate-perception range.

The consequences of six-day (or five-day) weeks filling up regular months in regular years, you ask? Oh, boy. Well, union contracts would have to be rewritten, TV broadcast schedules rearranged, birthdays and anniversaries would always fall on the same day of the week ...

But the *biggest mathsemantic* consequence might be that some cultures would miss the decimal system. The example of *twelve* months in a year and *six* five-day weeks or five *six*-day weeks would have steered them to the duodecimal system. We might count by twelves. Zip, zap.

Don't be surprised. Twelve makes more sense than ten, fingers aside. Twelve divides up more ways. That's why groceries and other package-goods get sold that way.

The Babylonians even adopted a sixty system. Sixty happens to be both one sixth of three-hundred-sixty and the smallest number divisible by both ten and twelve. The Babylonians were strongly influenced by our "what if" world. That influence continues today in geometry and chronometry. We still count out our circles and our world into three-hundred-sixty degrees, our hours into sixty minutes, and our minutes into sixty seconds.

Now, actual moon phases recur in *about* twenty-nine days, twelve hours, and forty-four minutes, and actual earth orbits in *about* three hundred sixty-five days, five hours, and forty-nine minutes. That means an earth-year runs *about* twelve and seven-nineteenths lunar-months. Highly irregular.

Human desire for logic and simplicity, however, runs strong. To make apparent order from chaos, we humans can fudge the score with arbitrary mathsemantic conventions. That's what the ancient Egyptians and the revolutionary French did with their three-hundred-sixty-day calendar. They made it work by putting five or six leftover days at the end of every year.

Fudging on this scale takes specialists in occult science, historically a class of educated astronomer-priests. But they would have had less to do in our "what if" world. So its biggest consequence might have been societies with fewer class distinctions!

In my consulting work, I've gained real advantages from analyzing time mathsemantically.

The longest index listing in my 1981 book, *The Semantics of Air Passenger Transportation*, is, inevitably, "passengers." But the second longest is "time." Its subtopics show why: time cycles, time dimensions of travel, time in English and Hopi, ground time, time not in QSI (a government-sponsored Quality of Service Index), time of schedules, spatial metaphors for time, types of time, and value of time.

My strongest time-inspiration came from Whorf's contrasting Hopi time-semantics with English, which Whorf says follows the standard average European (SAE) subject-predicate pattern.

Korzybski's general semantics had softened me up for the differences Whorf described. Time is a difficult subject. St. Augustine put it this way: "For so it is Oh Lord, My God, I measure it but what it is I measure I do not know." Lewis A. Coser said in his presidential address to the American Sociological Association, "In fairness to St. Augustine, modern physics tends to agree with his position."

It's easy to get lost thinking about time. But the combination of Korzybski's insistence that—contrary to SAE implications—we can never know what something "is," Piaget's showing that children confuse words with things, and Bridgman's operational definitions, helped me battle SAE presumptions to get at the Hopi approach.

Piaget's assumption that children move to adult formulations means to adult SAE formulations. Perhaps some people don't move far enough. That's one problem. But, as Korzybski argued, perhaps the adult SAE view itself is inadequate. That's both a problem and a business opportunity. So I had incentive.

In English, we objectify time. We count days and report the result in cardinal numbers: *three* days. The Hopi would treat such an imaginary plural as illegitimate. They'd say they can see three sheep together, but not three days.

Indeed, one can experience three days only sequentially, not all at once. Therefore, the Hopi sequence days and report the result in ordinal numbers: the *third* day.

We visit and stay three days. They visit and leave after the third day. We say, "Sixty days is longer than thirty days." They say, "The sixtieth day is later than the thirtieth."

Not content merely to objectify time, we then dress it in spatial metaphors. We speak of "points" in time, "lost" days, and "big" moments. We use words like "before," "after," "long," and "short," for both space and time. We even apply spatial metaphors to cyclical events. We speak of "heavy travelers" and mean not overweight passengers but just people who travel frequently. We speak of a "big market" and mean one in which many people travel frequently.

English permits us to say "Tomorrow is another day," as if turning to a clean page in a blank book. Hopi doesn't do this. It conditions its speakers—properly, in my view—to regard tomorrow as a continuation of all that has come before.

Our language doesn't distinguish between two events taking place at the same time, such as two TV shows, and two events that never can take place at the same time, such as two Saturdays. Therefore, we lose track of the distinction.

This mathsemantic difference has powerful repercussions. It gave me a cognitive advantage. It underlies all the difficulties I've chronicled in counting passengers, customers, patients, callers, students, and missing children.

The other day on "Entertainment Tonight," John Tesh said that one hundred seventy million people had seen the 1970 movie *The Godfather* or its sequel. My wife and I groaned in unison. We presumed Tesh was reporting attendance, with its many repeat customers, not people.

Looking at time sequentially, as Whorf says the Hopi do, has also led me to see that better air service doesn't mean more flights. It means service better adapted to a person's time, to human diurnal, weekly, and other cycles. Our hours aren't equal. If anyone wants us to use a four o'clock flight, it had better be in the afternoon, not the morning.

"The longer it takes to get somewhere," my rule says, "the longer one must stay to make it worth the effort." Given instantaneous costless teleportation at will, we'd go all over the world and still return home to grab lunch, to check the mail, to sleep in our own beds. If you think about it, you can see that air transportation is simply today's fastest mode, and still truly inconvenient.

Studying time with a Whorfian mathsemantic twist has helped me see that changes in passenger counts generally reflect changes in travel-time dimensions. If service improves, total-days-away may change surprisingly little, yet passenger counts increase dramatically because more-trips-of-shorter-duration replace fewer-trips-of-longer-duration. Quantifying these reciprocal changes in time dimensions has put a lot of meals on my table.

There's even an ethical side to time-semantics.

Korzybski defined human beings as the time-binding class of life. Plants, he said, are mostly engaged in chemistry-binding, to which animals add space-binding. But people excel in transmitting information through time. We use systems developed by the ancient Babylonians, Egyptians, and Greeks. Korzybski felt that human progress depended on using time-binding properly. It's an interesting thought. For example, it puts a premium on clarity. It puts a premium on setting a good example. It makes you think about those who come after.

The ancient Greeks, says Oswald Spengler, lacked an adequate sense of time. The Egyptians had it; they dated events, they contemplated eternity. The Greeks didn't; they cremated their dead, they stuck to the here and now. Spengler calls Greek attempts at framing calendars "naive." He cites a Greek treaty that proclaimed itself "valid for a hundred years from this year" without even identifying what "this year" was. Strange, isn't it? We, in the West at least, look more to the Greeks than the Egyptians for models.

Thinking about time also brings one face to face with these civilizations that have perished. What happened to the peoples of those ancient times who thought they had it made? Will that happen to us?

Finally, time-semantics played the leading role in what is arguably the twentieth century's greatest intellectual revolution. How many people know that Einstein's relativity theory grapples with the problem of setting clocks? Or what kind of a problem this is? "Einstein's *relativity of time,*" wrote Philipp Frank, "is a reform in *semantics,* not in metaphysics." (His emphasis.)

And here's a marvelous insight into two marvelous people. Piaget's researches into children's conceptions of time were suggested by Einstein. The physicist, who was presiding at a 1928 symposium on the philosophy of science, became interested in Piaget's report on children's thinking and inspired him to look into how children's thinking related time and speed.

Which suggests that mathematical physicists can learn from child psychologists and vice versa. And maybe that we've still got a lot to learn.

A friend once cautioned that if I ever felt the urge to be poetic, I shouldn't be original. Therefore, I'll leave you with a proposition about time, the human condition, and mathsemantics that pays homage in a way to another Marvell. Andrew.

Proposition 23: Had we but world enough, and time,
Bad mathsemantics were no crime.

CHAPTER 19

One for all

Inow need to tell you that all the quiz examples given so far come
from just one of two recruitment quizzes we used. I haven't mentioned the second quiz earlier for fear of adding complications that
don't matter.

The second quiz was addressed to one hundred ninety-nine analysts
and administrative-law applicants. It contained some questions presumably more difficult than those in the first quiz.

One of these questions is reproduced below. The emphasis shown is
in the original.

> Would it affect the <u>number</u> of trips of 500 miles or more
> which would be made, if the local airport were 10 miles far-
> ther away from the prospective passengers? (Yes or
> No) _____
> Why? _____

As you can see, the question has two parts. The first part calls for a
simple yes or no. We received essentially four different answers.

The question's second part asks applicants for reasons supporting

their answers to the first part. Except for seventeen blanks, no two of
these answers were exactly the same.

One hundred sixty-two applicants, more than four of five, answered
"no" to the first part, thereby signifying that a more distant airport
would not affect the number of passengers. This is a remarkable show
of agreement. Unfortunately, the answer is wrong.

Nine applicants, one in twenty-two, gave no answer.

One applicant answered, "Yes, in a positive way," as if to argue that
more people would use a more distant airport. This reversal of the principle of least effort falls into a category by itself.

Only twenty-seven applicants, just over one in eight, gave the correct answer, "yes." Their reasons strike many nails on the head. "People are basically lazy," lands hard on the theory of least effort. "Cost
of using airplane goes up," hits on the economic principle that when
cost goes up, consumption goes down. "People, once having been irritated by the 10 miles, may think twice before repeating," pounds the
recycling principle, one person as many passengers. "Fewer trips
could be arranged to spend the same amount of time away," then
drives home our choice of reciprocal changes in the number and duration of trips.

Don't be concerned if you, personally, aren't yet convinced. It might
even increase your interest in the real question here. That question is,
"Why did more than four out of five applicants get this question
wrong?"

The answer is the single-instance habit.

The single-instance habit is my name for the way we so often automatically reduce many instances to a stereotype, a prototype, a best example.

Our basic cognitive ability supporting the habit has worked well.
Categorizing and stereotyping—summarizing large striped feline images as "tiger" and "always dangerous"—helped our ancestors survive. Hence, given how evolution works, the single-instance habit
survived. We couldn't completely eliminate it, even if we tried, and I'm
not suggesting we try. I do suggest we use it more consciously.

Automatically reducing many instances to one has some unfortu-

nate effects. Perhaps the easiest to see are those arising from preju-
dice, from mean-spirited group stereotypes applied hurtfully to individuals.

But I want here to leave out questions of bias, for they tend to direct
attention to controversial social issues. I want to focus, at least initially,
on just the mathsemantics aspects of the single-instance habit.

The effect of airport distance on passenger traffic makes a good example. It illustrates a majority habit; four out of five applicants for
nonclerical jobs got it wrong. It illustrates a habit that most educations
and experiences leave intact; virtually all who were asked the question
were college-trained and already working. It illustrates the danger of
reducing large numbers to single instances. It's a matter to which most
can relate. Yet it hasn't inspired any protest marches.

Single-instance reasoning showed up in the quiz answers, sometimes very clearly. "A distance of ten miles is not enough to deter
someone from making a 500 mile flight." Note the reference to "a"
trip, one trip. "Because in a trip of that distance the 10 mi would not
mean much. The alternatives are worse." "For trip of 500, passenger
will probably be willing to travel extra 10 miles." "Once the decision to
travel is made the distance to the a/p [airport] of 10 miles would not
be a great enough deterrent to have much effect." "In my opinion, a
10 minute longer drive to the airport would hardly stack up against an
8 1/2 hour drive to one's destination." "The marginal effort to travel
the extra 10 mi. would be insignificant compared to the total time
saved by flying 500 miles."

Single-instance reasoning came out even more pointedly in discussions after the quiz.

"If I had to make a five-hundred-mile trip," applicants would
argue, "then ten additional miles to the airport wouldn't make any
difference."

"But," I'd ask, "wouldn't it raise the cost?"

"Not much," they'd reply, "when you consider the whole trip."

"But," I would then continue with my best economic argument,
"when costs go up, consumption comes down."

"Not in this case," they'd argue, picking up the economic thread, "because there isn't any good substitute for flying on trips that long. Driving takes too long."

"How about the telephone?" I'd suggest.

"If the telephone would do, I'd have used it in the first place."

"Then, let me try a different approach. You said 'If I had to make a five-hundred-mile trip.' Those were your words. That assumes you're going to make a trip, doesn't it?"

"Yes, but I'm not sure what you're driving at."

"Only this: The question was whether there would be a change in the *number* of passengers, or trips, and you're trying to get the answer by analyzing *one* trip that *has* to be made."

"So?"

"I don't think we can study change by assuming it away. Your assumption says there won't be a change. One trip that has to be made remains one trip, so you're not really reasoning out an answer. You're just assuming it."

"Well, I've got to think about a trip, don't I? How else could you think about the problem?"

And then I'd have them. "Well," I'd say, "before we can reason about this question, we must drop the assumption that there's only one trip. I suggest we think, say, about a million people who would otherwise make three million round trips a year generating a count of six million passengers, except that the air service has been moved to another airport ten miles farther away. Now let's ask, 'Would we lose any traffic?' "

"Oh, well," most would now concede, "if you look at it that way, I guess a few trips would be lost, but nothing significant. The practical result is still that you wouldn't lose any traffic to speak of."

I loved this opening. Its narrowness illustrated the hold of the single-instance mathsemantic habit. But an opening is an opening, and I had a crowbar.

"Would you say that losing one passenger in ten was a significant loss?"

"Sure, but I can't believe that."

And then I'd show them the evidence. As many studies as they liked. From different points of view. All pointing to the same conclusion.

The extent to which people repack their travel into fewer trips of longer duration when travel costs go up surprises almost everybody. The combined lesson of studies over thirty years is that each added mile to an airport reduces passenger traffic by about one in a hundred. Ten added miles reduces traffic by about one passenger in ten.

This is a surprise, as I just said, for almost everybody, and difficult for many to accept even when they review the evidence. They need reasons that cut through, that destroy, their mathsemantic many-into-one fallacies.

The first many-into-one fallacy is that a passenger (the many) is a person (the one). Remember the year that I became ninety-four passengers.

The second fallacy is that the factors that influence a series of trips (the many) can be analyzed by studying a single trip (the one).

The evidence made applicants uneasy.

"Do you feel," I'd ask, "that losing one in ten passengers is too many?"

"I sure do. How could that happen? How can anybody cut that many trips?"

"Good question," I'd agree, "but say you had to spend twelve days each year in a branch plant, which you were in the habit of packaging into four trips of three days each. By repackaging the twelve days into three trips of four days each, you'd cut your trips by one in four. That's much more than one in ten."

"Are you saying that's what happens?"

"Well, yes and no. It's only one example of the sort of thing that happens. Only a few would revise their travel practices that much. Most people wouldn't. But then, for a few people, the extra airport distance would be the last straw convincing them to accept a job in another city, so we'd lose all their trips. I'm only saying that when you look at lots of trips being made over time by few people, you can see how losing

one trip in ten isn't that far out. I can show you much worse losses relating to other changes in service convenience."

That usually did it. Occasionally I'd then get the argument that this
could make sense for short-haul trips, but not for long-haul ones, and
I'd realize my performance hadn't overcome the single-instance habit.

Contrary to expectations based on single-trip thinking, the passenger losses caused by more distant airports (and service inconvenience
in general) occur in all mileage bands. Once beyond three hundred
miles, each band loses about the same ratio of its total trips as any
other band. Exactly why is unknown. We do know that any extra travel
time pushes some long-haul trips over into another day, providing an
extra incentive to repackage long-haul travel into fewer trips of longer
duration.

Apart from single-instance reasoning, we have no reason to be surprised by this. But single-instance reasoning seems often to provide
the initial impression that a more careful analysis must then struggle
to overcome.

Prior to my 1981 book, most engineers and analysts argued, with
scant evidence, that airport location didn't significantly affect traffic.
Apart from a brief revival in Canada by supporters of Montreal's distant Mirabel airport, I haven't heard that argument since publication.

The extent to which people deny that convenience affects passenger
traffic should dismay economists who teach "higher price, lower consumption" as a basic principle. Many of our applicants had studied economics, yet readily abandoned the principle in favor of single-instance
reasoning, apparently the stronger mathsemantic habit.

The airport question concerns our own decisions relative to simple
changes in ground travel. There are no great mysteries here, no arcane statistics, just ordinary time and space. Everyone has a reasonable idea what ten miles means in terms of ground travel. Yet
well-trained applicants miss the call.

I believe that we must conclude that our natural inclination to reason
from one instance can disable our thinking regarding simple repetitive
behaviors.

What's scary is this: If we can't make reasonable judgments about simple repetitive matters involving ordinary space and time, what does that suggest for our decisions about more complicated things?

Professors Slovic, Fischhoff, and Lichtenstein take up the case of automobile seat belts.

> Research has demonstrated that seat belts are effective in reducing death and injury in automobile accidents and that most people are aware of this fact. However, the percentage of motorists who wear them is small, and numerous and expensive media campaigns have failed to persuade people to "buckle up for safety."

Does this mean we're stupid?

It depends, say the three authors, on how the risks of accidents are framed. "Because a fatal accident occurs only once in every 3.5 million person-trips and a disabling injury only once in every 100,000 person-trips, refusing to buckle one's seat belt prior to a single trip is not unreasonable."

However, they say, the risks can be stated differently. Over the course of a fifty-year lifetime of driving, the average motorist has one chance in a hundred of suffering a fatal accident and one chance in three of suffering a disabling injury.

The authors say that exposure to the single-trip statistics have far less influence on seat-belt use than the lifetime statistics. I can believe it.

Left to their own devices, however, most people act as if they had been exposed only to the single-trip statistics. This suggests that reducing multiple instances to a single example can be a particularly dangerous mathsemantic habit.

The effects of many events, such as littering, giving birth, and cheating, to name just three, cannot be captured properly in a single instance. If we base our reasoning on just one instance, we miss the effect of repetitions.

Single-instance reasoning seems to underlie a problem Paulos puts so differently that we could easily miss the connection.

> One important [factor contributing to innumeracy] is the impersonality of mathematics. Some people personalize events excessively, resisting an external perspective, and since numbers and an impersonal view of the world are intimately related, this resistance contributes to an almost willful innumeracy.

I agree that some people don't look much beyond themselves. I also agree that numbers can help foster a more general view of the world. However, I hate to call it an "impersonal view." That smacks too much of the "rational man."

I don't take numbers impersonally. Quite the contrary. I'm strongly disposed not *ever* to get into a fatal or disabling automobile accident. The statistics get to me personally, not just externally or impersonally.

If we want people to think in terms of multiple instances at large, however, shouldn't we at least ask first whether they can think in terms of multiple ones at home? Shouldn't we address the single-instance habit before we get into personalities?

Talk about "math's personality" could lead us astray. A language isn't a person, obviously, so speaking of a language's "personality" is a kind of shorthand that takes deciphering. If a language can have or reflect a personality, then presumably it would be the personality of its creators or users. I agree with Whorf that a language influences its users, although I don't know how far. I'm certainly willing to grant that we can use math to get a more general view of what's going on. But I also believe one can become obsessed with numbers and use them with no more perspective than Dustin Hoffman as "Rainman" counting toothpicks or a teacher demanding "right" answers.

I believe we should address mathsemantics rather than issues of personality versus impersonality. I find the habit of reducing recurring events to single instances no more and no less necessarily personal than my habit of insisting on viewing multiple instances. Therefore, I argue in favor of viewing multiple instances, not as more "impersonal" but as more useful.

This chapter criticizes the habit of automatically reducing recurring events to single instances as a bad personal habit to be replaced by a better personal habit.

This isn't a purely individual matter. Like many other bad habits, the single-instance habit has its institutional supports.

Communication specialists teach quite correctly that a single well-told instance has great power to move people. The three gray whales trapped by oncoming winter ice off northshore Alaska late in 1988 got more media attention and immediate action than more critical cetacean problems. Some whales get trapped by Arctic ice most years with little effect on overall whale populations. Yet this story was a cliff-hanger, a drama of the kind we all understand from millions of years of evolutionary survival that had little help from mathsemantics.

Some news commentaries used the plight of the three gray whales, as I do here, to point up a bigger story. Good for them. There's a limit to what media can do in changing habits. That's not their primary job. I'd just like to see some emphasis somewhere in the communication industry on the ultimate dangers of encouraging and milking the single-instance habit.

Averages seek to retain the power of the habit by reducing a multitude of phenomena to single composite instances. It's a brave attempt. When the events being combined don't vary too greatly, say, the number of calories in a slice of bread, we can all benefit from using averages. Yet their danger is so obvious it's even the subject of a math one-liner about drowning in a river that averages only three inches deep.

I remember hearing a national call—President Roosevelt's, I believe, sometime around 1940—to extend the draft age down to eighteen *to reduce the army's average age* and joking at the time to myself that a faster way would be to draft infants.

The *New York Times* of February 28, 1989, prompted by United's losing a cargo door over the Pacific and Aloha's losing part of a cabin roof in flight earlier in the year, ran a first-page story on airline safety. They tied the accidents to aircraft aging and included a table showing the average age of planes in each major carrier's fleet.

In the absence of other information, knowing the average may be better than knowing nothing, but the crux of the matter is how many planes are nearing a dangerous age. The article stated that planes are designed to last about thirty years, and I would then have liked to see a table showing me what share of each carrier's fleet was over twenty years or otherwise nearing thirty.

Sometimes the single instance isn't what matters, but only a *cumulative effect*. Take oil company highway rest rooms. You expect a dirty one once in a while. What really gets you is several dirty ones in a row. A survey that compares companies shouldn't just say what percentage of the rest rooms are clean. It should give your chances of hitting, say, three dirty ones in a row. If nineteen out of twenty of Company A's rest rooms are clean, then your chances of happening on three dirty ones in a row are only one in eight thousand.

If three out of four of Company B's rest rooms are clean, your chances of happening on three dirty ones in a row jump to one in sixty-four. This is a one-hundred-twenty-five-times greater risk than with Company A.

And if, for the sake of illustration, the magic number at which customers switch brands is three dirty rest rooms in a row, then Company B will lose one hundred twenty-five customers for this reason for every customer lost by Company A. One wouldn't have guessed that from the single-instance difference.

Single-instance stereotypes enter into all kinds of serious debates.

For example, union officials and other supporters argue that without minimum wages, employers would take advantage of the workers most in need of jobs and least able to defend themselves. The picture we receive is that increasing the minimum wage helps single-parent-household mothers.

Laissez-faire economists opposed to minimum-wage laws, however, emphasize that requiring wages higher than the market rate means that marginal workers, especially inexperienced minority youths, will not be hired. The different picture we receive here is that increasing the minimum wage will cost black teenagers their jobs.

Charles Brown of the University of Michigan analyzed this controversy in terms of many statistical studies and concluded that "the minimum wage is overrated: by its critics as well as its supporters." Two of his subsidiary points are especially pertinent here.

The first, which I didn't expect, is how few low-paid workers are members of low-income families. Nine out of ten belong to families higher on the scale. That would mean a low-paid worker is seldom a family's major breadwinner.

This aroused my curiosity, so I consulted the *Statistical Abstract of the United States*. I found that of 3,927,000 workers in 1988 at or below the minimum wage, 3,236,000 were white, 3,207,000 worked in service industries, 2,615,000 were part-time workers, 2,550,000 were women, and 2,281,000 were sixteen to twenty-four years old. That would seem to make the prototypical minimum-wage worker, a white, young, female, part-time, service-industry worker from a multiple-worker family.

Into my mind jumped the image of a neighbor's daughter dispensing hamburgers after school at a local fast-food franchise. The image pained me as incongruous, unreasonable, unfair, unsympathetic, a narrow-minded prototype. I fought it off.

The second point that surprised me in Brown's article is that minimum wages have had little effect on employment, even among black teens. I've no quick way to check that point.

Don't expect me to offer a final judgment on this controversial subject. I'm not an expert on wage laws. If Brown is right, however, the data support neither a stereotypical argument for, nor a stereotypical argument against, minimum wages. Whatever effect minimum-wage laws have on low-income families is apparently small overall— probably important in specific cases but mixed in with effects on more affluent families. The picture isn't very clear.

Indeed, our understandings of most employment and wage issues are fraught with mathsemantic difficulties. What does unemployment mean? No one has yet defined it satisfactorily. Completely discouraged persons who stop looking for jobs are not counted among the "unemployed," because they have removed themselves from the labor market. What are "wages"? Do they include frequent-flyer benefits earned on company business but used for personal travel? Currently no. But stay tuned. And what about the statistics? Do individuals and companies tell the whole truth about employment and wages? You tell me.

Then, what about the problems of double counting over time, people holding multiple jobs, changing jobs, being hired in the same industry twice in the same year? Brown remarks, "One should be careful about contrasting 'those lucky enough to work at a higher wage' with 'those who are unable to find work'; with high turnover rates, the two groups *may* be composed of largely the same people at different moments."

I don't bring up these matters to make social or economic points; I don't know enough to do that. I bring them up to illustrate the immediate point that reasoning from a single instance is dangerous, and the more general point that mathsemantic problems plague our understanding.

In December 1988 over Lockerbie, Scotland, less than one pound of Semtex destroyed a Pan Am 747 and its two hundred seventy passengers. What can we do about bomb threats?

Note how the question refers to a single instance when we look back but to innumerable instances when we look forward. This common mathsemantic metamorphosis helps to explain why complaining is so much easier than finding remedies.

Retrospectively, we can always conclude that more safety precautions would have benefited those who suffered from a specific disaster. Thinking in terms of a single instance, we feel there must be a way to detect and avoid plastic explosives in checked luggage.

Looking forward, however, that could mean checking more than a billion (1,000,000,000) bags each year. Neither people nor machines work perfectly. Therefore, we must figure that any practical system would miss some real bombs and set off some false alarms. For any given technology, the surer we want to be, the more false alarms and delays we have to accept. Somewhere—it's hard to know exactly where—the costs and delays associated with improving safety out-

weigh what that degree of safety is worth for the system. Given another disaster, do we tighten the system? Or do we find it falls within
the acceptable risk?

Looking at one instance can fail to capture what we face in the world.
If we are to make wise decisions, we must avoid this mathsemantic
fallacy.

Proposition 24: Collapsing multiple instances into a single
one can be a particularly risky mathsemantic habit.

CHAPTER

Estimating

"When," I once asked a job applicant, "do you think the capital
of Burma was founded?"

"I've no idea."

"But," I persisted, "you must have some idea."

"No, really, I have no idea at all. I don't even know what the capital
of Burma is. Is it Rangoon?"

"Sorry," I parried, "but the question isn't 'What is the capital of
Burma?' It's 'When do you think it was founded?'"

"Well," came the reply, "I can't say, if I don't even know what city it
is, can I? I just have no idea."

"Please," I challenged, "stop giving me false information about
yourself. You must have some idea, and I'm asking you for it."

"You want me to guess? Is that it?"

"Not at all," said I. "I want you to use what you know to make an
estimate."

"But I don't know anything about Burma," the applicant protested.

S. I. Hayakawa, who once taught English, tells of discovering among
the papers handed in at the end of a classroom writing assignment one
sheet bearing nothing but the student's name, the date, and a hole

worn by successive erasures of false starts. The future senator concluded that a common residue of English instruction is an abiding fear
of using one's own language.

Math instruction seems to have a similar effect. First, as Kamii
makes clear, we teach children to distrust their own thinking by making them use unnatural math algorithms. Then we teach them that
small errors are considered errors just like big ones. No wonder we
learn to play it safe, and wait to be told, rather than exercising a little
intellectual autonomy. Why take the chance?

What I try to do with my Burma question and others like it is find
out whether applicants are willing to make estimates in challenging situations. Some will. Most won't. They fear being wrong.

That fear discourages what is probably our least expensive and most
valuable learning technique, trial and error. Done right, cognitive trial
and error costs next to nothing. We need only make, and then check,
estimates of what's around the next corner and under the next stone.

A few hours after writing the foregoing I was enjoying a cup of coffee with my wife and took the occasion to ask her, "When do you think
the capital of Burma was founded?"

"Do you want me to answer," she asked, "even if I don't know the
capital of Burma?"

"Yes."

She thought a moment before answering.

"About 1300."

Then I told her what I'd been writing.

You have to realize that my wife screened most of our applicants, ad-
ministered most of the quizzes, and often knows what I'm about to say
before I do. Her 1300 sounded good to me.

Then, to satisfy my curiosity, I looked it up. According to my source,
the capital is Rangoon, probably founded in the sixth century, and until
the eighteenth century a small fishing village dominated by Burma's
most celebrated temple. It was named the capital in 1753.

In our immediate family no one gets credit for knowing anything unless it's stated before the proof, but everybody gets credit, bravery
credit, for voicing an estimate.

"Oh, yes," you might volunteer, "I knew that."

If you did say that, then by our rules I'd have to respond with something like, "Sorry, dear reader, much as I would like to, I can't give you any credit."

The point of our family practice is not to doubt anyone's honesty. The point is to provide incentive for taking chances openly.

To become good estimators, we need most of all to (1) make lots of checkable estimates, and (2) treat knowledge as a web.

Fear usually stops people from making checkable estimates. If we need to know when the capital of Burma was founded, say, for a parlor game or a school paper, we typically just go look it up. Then we use whatever we find.

Having made no estimate, we can't be wrong. Most people would see this as an advantage.

I've gone so far down the other track, I see such reticence as a disadvantage. Oh, I used to be afraid to make estimates. In my first consulting job I marveled at how our firm's principal stuck his neck out making forecasts and recommendations. I didn't think I'd ever know enough to be able to do that.

What I've learned is that no matter how much one knows, one can't make forecasts without a willingness to be proven wrong. There's no way to be sure what's going to happen. No amount of information will ever be enough, for there isn't any sure information about the future, only estimates.

In one sense, every estimate is about the future.

"Hold on a minute," you might say. "Your wife's estimate that the Burmese capital was founded in 1300 was about the past, wasn't it?"

Yes, of course, you'd be right, but it was also about the future, about what we'd find looking it up. The date was a checkable estimate. The sense in which I mean that every estimate is about the future is that the checking is in the future.

The fastest way to learn to make estimates is to make lots of them that can be checked right away.

For example, let's say you're cruising country lanes with a local surveyor to learn how to estimate acreages. No matter where you start,

MATHSEMANTICS

no matter how far off you are, if you get immediate checks, your estimates must improve rapidly. What difference does it make whether you estimate that first ten-acre field at one acre or ten thousand? As soon as you learn it's ten acres, your next estimate will be much better. Just three or four estimates later, you'll be into the fine points.

But that first estimate is the mathsemantic killer, isn't it? You fear you're totally ignorant, about to make a fool of yourself. You've grown up entirely in the city, perhaps, and are convinced you have no idea what an acre is, that you wouldn't recognize one if you stepped on it.

This is where treating knowledge as a web comes in. You've heard of "acres." It's not a new term. Even a city dweller must have heard or read it somewhere. You probably know it's a way of measuring land area. You must have heard of suburban lots, farms, or ranches being measured in acres. If asked, you'd probably say those acreages weren't in hundredths, nor in millions, but in amounts like three, a hundred, or a thousand. So, even as a city dweller, you might know as much about acreages as my wife did about Burma's capital.

Spiders spin webs to catch insects. Humans connect strands into nets that catch hummingbirds, lions, and tuna. Most of any web or net is open space.

If the strands weren't connected, they wouldn't make a web. They'd just be loose bits and pieces of line.

A lot of people treat their knowledge like that, as just so many disconnected facts. They think they can't estimate acreage or dates because they lack the particular facts.

The path to better estimating lies in treating your knowledge as a web. Imagine everything you know as having lines connecting it to everything else that has ever been or could ever be. You can't see the lines, but they must be there; for nothing exists in isolation. Everything you know is a clue to a great deal more.

The more clues you connect, the stronger your web.

The farther out the clues are, the larger the web. Even a loose mesh, like your knowledge of dates, Asia, and cities, will enable you to make rough estimates of interior points, like the founding date of Burma's capital. You wouldn't estimate any city at one million B.C., would you?

The closer together the clues are, the smaller and tighter the web. A lot of clues close together makes a tight web of guaranteed but limited utility, like a kitchen strainer.

A good arrangement is a small, tight web within a gigantic open one. The small web gives you a home, a specialty, your area of expertise. The great loose web guarantees that you'll seldom be very lost for very long.

I remarked in an earlier chapter that the single most useful number for a marketing analyst in the United States is our total population, not just for itself but for the bearing it has on almost every other number. In terms of the web analogy, it's like a point with strands going in surprisingly many directions.

The February 1989 issue of *Soviet Life* had a question-and-answer page titled "Readers Want to Know." Its first question was, "How many marriages are performed every year in the Soviet Union?" The answer given was, "Every year about 2,700 couples get married."

Don't believe it.

Let's use my favorite starting point, the total population of the United States, about two hundred fifty million (250,000,000) people. One of the facts in my web is that the (now former) Soviet Union had somewhat more people than the United States. Another is that our average life span is a little over seventy years. I'd estimate it can't have been too different in the Soviet Union. Another is that a marriage takes two people, which I presume is pretty much the same everywhere. And still another is that most people get married, some more than once. If, for mathematical convenience, I take the population of the former Soviet Union as two hundred eighty million and their average life span as seventy years, then I can figure the average number of people reaching any given age in a year as two hundred eighty million (280,000,000) divided by seventy (70). That comes out to four million (4,000,000). Dividing by two gives me two million (2,000,000) "couples" coming of age each year.

But if most people get married, then the chance that *only* two thousand seven hundred (2,700) out of two million couples will get married is, for all practical purposes, zero. It's much likelier that somebody at

Soviet Life, just like some of our applicants, didn't know millions from thousands and dropped three zeros from the correct number, two million seven hundred thousand (2,700,000).

For fun, I looked up the number of families and marriages in the United States, which for 1988 was 65,133,000 families and 2,389,000 marriages. Compare that with 70,000,000 families (another *Soviet Life* figure) and 2,700,000 marriages for the Soviet Union. Looks good, doesn't it?

If I absorbed only discrete "facts," I might have accepted the figure of 2,700 marriages as coming from a "good" source. How would I know it was wrong?

But my mathsemantic web automatically recognized it as probably wrong, even though the *number* of marriages anywhere, let alone in the Soviet Union, is nothing I'd ever thought about before reading the figure in *Soviet Life*.

No strand of the web used here is at all remarkable. The events considered numerically are the U.S. population, the Soviet Union's having more people, our living to about seventy years, that a marriage consists of two people, and that most people get married at least once.

The "pure" math includes knowing that twenty-eight can be divided evenly by seven, that two hundred eighty million divided by seventy years is four million per year, and that four million per year divided by two is two million "pairs" per year.

Hardly genius-level stuff, factually or mathematically. Just some webby mathsemantic know-how.

The *Philadelphia Inquirer* of July 8, 1989, in a story about mannequins designed to discourage shoplifting, reported:

> Shoplifting and employee theft cost the nation $508 billion annually, said Detective Gail Riddell, who heads the Denver Police Department shoplifting unit.

My mathsemantic web registered a sudden tug, as if to warn, "Something just landed. Look out." My web does that a lot. I don't have to connect it up or turn it on. It's a routinely functioning part of my cognitive apparatus.

My immediate in-the-head calculation, rounded off as always for my convenience, was that $500 billion theft annually divided by 250 million people (that number again) is two thousand dollars ($2,000) theft per person, infants and grannies included. "No way," I said to myself.

My next calculation depended on knowing not all of us steal. "Even if one person in ten were a store thief," I thought, "which must be too high, they'd have to average twenty thousand dollars apiece. That's absurd."

I looked up total U.S. retail sales in the *Statistical Abstract of the United States,* the citizen's mathsemantic book of books. The *highest* total retail sales figure I could find was just under $1,500 billion. That would make the reported $500 billion store theft equal to one third of total sales. I didn't believe that possible.

So I looked up robbery and property crimes. Same book, different table. It reported just under a million (983,000) shoplifting offenses averaging one hundred and four dollars ($104) each for 1988. A million times a hundred dollars has to be a hundred million dollars (actually $102,232,000 if we include the useless details) annually.

This shoplifting figure may exclude employee crime. It doesn't matter. The difference between $500 billion and $100 million seals the fate of the first figure. The difference is huge, $499.9 billion.

Got that last bit? It's pure math. Also, $500 billion is five hundred times $1 billion, right? So it's five *thousand* times $100 million. My brother and I didn't count that high in Dad's ledger-paper challenge, but that's where my hand and brain began feeling the huge difference three zeros can make.

Perhaps this is another case of somebody garbling billions and millions and not knowing the difference. That's being off by a factor of a thousand. It's an error equivalent to saying a one-year-old is a thousand years old. Would anybody stand for that? We seem to permit a double standard in communication, generally high except for mathsemantics, where almost anything goes.

Incidentally, the same table enables one to estimate that far fewer than one in a hundred of us shoplift. It doesn't report shoplifters, naturally, only shoplifting events, the old person-passenger problem. There's about one such event per year for every two hundred people. To the extent the average shoplifter accounts for more than one event, shoplifters will number fewer than one in every two hundred people, possibly far fewer.

While driving in Colorado one summer, I found myself in the mathsemantic web of an aviation specialist. My wife and I heard a radio news bulletin of a commuter airline crash that had killed twenty passengers. "Oh gee," I said to my wife, "there was a baby on board." "How do you know that?" she asked.

Sadly, it turned out that my surmise was correct. A baby had died. Here's how I had "known." The kind of commuter aircraft involved was certificated to hold a maximum of nineteen passengers. "Passenger," of course, doesn't mean "person." The federal regulation excludes infants not occupying their own seats. Stowaways and standees are exceeding rare. Hence, the newscaster's twenty "passengers" must include an infant.

Mathsemantic webs are personal. No two are exactly the same. Once in place, they work automatically, immediately, and privately. I've tried to tell you how my web feels as it works. You can't learn about mathsemantic webs from formal reports. Formal reports get cleaned up for presentation.

One needn't be a specialist to construct a mathsemantic web. You don't need to know much math. Just start using it. Start making numerical estimates of events. Check them. Repeat. And repeat.

Anybody can make many checkable estimates a day. For instance, while walking, estimate how many people you'll see around the next corner. While driving, estimate how far it will be on the odometer to where you're going. Before buying gas, estimate how many gallons it will take. Before looking at your watch, estimate what time it is. Before opening the envelope, estimate your electric bill. Before opening the book, estimate its total pages. Before looking at the thermometer, estimate the temperature. As you watch football, estimate each play's yardage. As you watch a movie, estimate when it was made.

For that matter, don't count anything, look up, or ask for any figure without estimating it first.

The difficult thing isn't finding opportunities to learn estimating, it's committing to an estimate. Checkable estimating opportunities come at you all day long. The trick is getting yourself to take advantage of them.

You must say your estimate *aloud*, even if only to yourself. Don't just think it. That's not enough commitment. Thinking is too easily denied. Say it out loud.

And don't hedge. Pick *one* number. Just one. For example, "Three-yard gain." Not a range like "three or four." Not a limit like "less than five." Just one number. Put yourself on the line.

Of course, most of the time from the "pure" math point of view you'll be wrong. Shrug it off as irrelevant—it really is. Virtually all of my estimates, even in my specialty, when viewed from a "pure" math point of view are wrong.

Mathsemantically, however, it's a different story. Your estimates will almost surely get better up to a point. When they cease to improve, you'll get a sense of how far off they can be and under what conditions. You'll spin yourself a bigger and better mathsemantic web.

As it grows and improves, you'll find your web working as a safety net, quietly alerting you to what's going on.

You'll find you've developed a different kind of memory and a different attitude toward memorizing. You'll worry less about specific numbers and take increasing comfort from being able to make adequate estimates as needed. You'll look for key numbers, like the U.S. population, to which you can attach lines running in all directions. You'll learn a few simple mathematical relationships, like a million being a thousand times a thousand.

Next, you'll find yourself extending your range, noticing and estimating events beyond your daily routine. For example, within days of drafting this chapter, I learned that the names "Rangoon" and

"Burma" were changed in 1989 to "Yangon" and "Myanmar." My net suggested "Yangon" might be closer to the local pronunciation, apparently correct. However, it suggested nothing about "Myanmar." Now I know that the Burmese are only one of several ethnic groups in the country, so the change was made to satisfy those formerly left out, much as "mailman" became "letter carrier," and "stewardess" became "flight attendant." You can't be sure what a web will trap, but the more you use it, the better it gets.

Similarly, as your web grows, it could well increase your ability to estimate events in business, in the news, in history, in national finance, in social statistics, in politics, in environmental matters, in almost anything you look into.

It should also extend your mathematical range, allowing you to feel effortlessly that a billion is a thousand times a million and sense more immediately when each makes sense.

End of motivational message.

Proposition 25: To develop your own useful mathsemantic web, estimate checkable figures out loud.

CHAPTER

On the fence

From 1948 to 1951, while I studied planning, my father introduced me as "commissar." Only after I'd received my degree and taken a job in private industry did he release me.

Perhaps Dad's sharp tongue was the key reason my parents separated and divorced. I don't know. It might have been money or sex, my father's book or my mother's singing. I really don't know. Maybe it was politics. I seem to remember Dad's hair turning white when Hoover lost his bid for reelection. On her own, after the separation, Mother put up a portrait of FDR.

Anyway, Dad said he was disappointed in me for studying planning at his old alma mater, the University of Chicago. "If you wanted to be a planner, son," he chided, "why didn't you go to the top, to the place where they know the most about it, to Moscow?"

He then said he was a Communist himself at heart, believing fervently in equality, freedom, a good living for all, the abolition of class distinctions, taking from each according to ability, and giving to each according to need. I'd learned not to be startled by anything my father said until he'd finished his point.

"Yes, son, I'd like to live to see that world. But," he continued quietly after a slight pause, "the Commies are kidding themselves. They're trying to get there the wrong way. They think progress can be planned from the center, that they can force the evolution of mankind. Everything we know about human incentives tells us it won't work. Their communism will turn out to be a step backwards, a denial of the freedoms and advantages we've gradually achieved starting with the Magna Carta. Worst of all, it'll deprive individuals of the knowledge and experience they need to make their own way forward, the only way we can get there, if we ever do." That was the end of the subject. We never spoke of it again.

I suspect that had he still been around, he wouldn't have been at all surprised by the world political events of 1989–90.

In one of my first planning classes, back home after WWII, the professor had us all cooperate in making an economic forecast. After briefly outlining the project, the professor called on each student in turn to suggest an assumption for use in the forecast. By the time he got to me, others had already mentioned such typical planning assumptions as that there'd be peace, moderately full employment, a low interest rate, no inflation, no revolutionary new energy source, and a post-war birth-rate surge followed by declining rates.

My turn. I bucked the trend. "Why don't we assume the world will be the way it actually is?"

The floor didn't open up to swallow me. My suggestion was too outlandish to merit comment. The professor went to the next student.

It took me three years to get my master's in planning. I wasn't slow, just going at the normal pace. Chicago back then under Robert Maynard Hutchins required undergraduates to take only general survey courses. After passing two years' worth of exams, you entered graduate school. I entered the Division of the Social Sciences and majored in something called the Program of Education and Research in Planning. The interdisciplinary aspect and the quality of the teaching appealed to me. I found myself simultaneously studying planning under New Deal Brain Truster Rexford Guy Tugwell, demography under Philip Hauser, economics under Milton Friedman, and sociology under Herbert Blumer. Fascinating.

I discovered that planners like to work with population and economic assumptions on which they can then build mathematical constructs called projections. It's less messy than taking the world as it is.
Planners then use the projections as if they were forecasts. However,
if things go awry and the planners are charged with error, they have
a refuge. "Now look here," they can counter, "we never said things
would turn out that way. All we said was that those base data and assumptions generate these projections. See, right here. It's all spelled
out in our report."

What we were being taught was how to make projections that
couldn't be proven wrong. The trick was to make enough assumptions and then use math properly. For example, assume that the
population of a city is one million and assume that it'll grow at one
percent per year compounded. Then in ten years the population of the
city will be one million one hundred four thousand six hundred and
twenty.

There's no way this "projection" can be wrong. If you make those
assumptions, that's the result you get. Every time. You can look it up
in a compound interest table. It's pure symbol system, pure math.

However, the "projection" sounds like demography. Most people
will think it's about population, about the world, that it's a "forecast,"
and that it can be right or wrong.

A population projection is pure math dressed up to look like a demo-
graphic estimate. In general, a projection of anything is pure math
masquerading as a forecast. It takes advantage of the childhood confusion of symbols with events. You might say it's on the fence between
pure and applied math.

Such projections leave room for unscrupulous use. That worried me
far less at Chicago, where everyone seemed honest enough, than the
fact that we weren't learning how to make forecasts.

As I attempted to show in the previous chapter, *learning* to estimate
requires a willingness to be proved wrong. A real forecast should contain estimates. Therefore, a real forecast can be wrong.

But we were learning to make projections that contained no estimates, only assumptions and math. We were learning how to make

plans based on projections whose internal logic was necessarily unassailable.

As a general semanticist, I felt we should try to be extensional, assume the world was the way it was, and make forecasts. Unassailable projections seemed somehow deceitful. I knew I was marching to a different drummer from the one my professors wanted me to hear. I didn't appreciate why.

I now see there's little incentive to teach students to do things that can be proven wrong. There's also little incentive to teach things that appear useless. Therefore, pedagogically, it makes sense to teach projecting and gloss over the distinction between projecting and forecasting.

That leaves one big problem, of course. How are we going to learn to make estimates and forecasts?

After graduation, I decided not to look for work in the public sector. I went into private consulting. That allowed me to assume the world is the way it is. I've never had to make assumptions and projections, only estimates and forecasts.

Before I graduated, Dad had made his generous offer to help me become an accountant, which, as mentioned earlier, I appreciated but turned down. Accounting sits on the symbol-system-event fence no less awkwardly than planning.

Prior to 1933, accountants certified the correctness of financial statements. That meant they stated officially that financial statements gave an accurate picture of events in the world, that they told the truth.

In 1933 Congress made accountants liable for false and misleading statements.

In 1934 the Federal Trade Commission countered by amending its regulations so that accountants need not certify a belief in the truth of their statements but only that they reflected the consistent application of generally accepted accounting principles. That's still the case today.

Accountants are not legally competent to value things. That duty rests with appraisers. Yet an accountant's figures seem to detail monetary values. A financier usually understands they both do and don't. But only an insider can know the extent to which they don't.

For example, a layperson thinks a balance sheet adds up all the assets, subtracts all the liabilities, and tells what a company is worth. That's not so. If it were, why would the figures be referred to as "book figures"?

What is a "book figure"? It's a figure in a company's financial books posted in accordance with accounting conventions. It could be a fleet of airplanes valued at cost less depreciation, regardless of what such planes could be bought or sold for. It could be a real estate parcel carried at its original cost in 1949 dollars, regardless of what it's now worth. It could be partially processed but unsold goods valued at the cost of the raw materials. It could be securities valued at cost rather than their higher but "unrealized" market value. It could be "good will," the presumed value of a business reputation. It could be imaginary vegetable oil valued as if it really were in the storage tanks. It could even be the presumed value of pharmaceuticals in imaginary warehouses.

Balance-sheet "book" figures differ from what a layperson might expect to the extent that accounting conventions differ from ordinary ideas of what things are worth.

Businessman Charles Ray Salmon has been an officer or director of more than ninety corporations. In his 1989 article, "The Balance Sheet: An American Fable," he comments on how well accounting figures correspond to events in the world.

> In more than 40 years of examining financial statements, I have yet to find one that does not contain misstatements, understatements, and other exaggerations. . . . Confusing numbers with reality appears to affect almost everyone, but some members of the accounting profession seem trapped in a digital dilemma.

The following writer agrees with Mr. Salmon.

> Business men would do well to familiarize themselves with this situation because it may affect them to a marked degree.

Present accounting methods may often compel them either
to overstate or to understate their earnings or their assets
and may practically prohibit them from publishing important
truths about their affairs. In some cases a certified public accountant's certificate can be obtained to the most flagrant
overstatement, while in other cases the accountant may insist
upon just as flagrant an understatement. . . . Hence, an honest management may be prevented from telling the truth, or
may perhaps even be prevented from knowing it, while a dis-
honest management may find itself in a position to take full
advantage of the distortion of facts.

That's a quote from *Truth in Accounting*, University of Pennsylvania
Press, 1939, reprinted in 1970 by Scholars Book Company as part of its
accounting classics series. The author was Kenneth MacNeal, C.P.A.,
my father. I believe he would have enjoyed Salmon's article. I can't be
sure. He was full of surprises.

Dad was an oddly caring man whose candor caused trouble. He just
couldn't or wouldn't see the emperor's clothes. So he said unpopular
things, in print, like: "The sincere and honest accountant of today is all
too likely to be the unconscious purveyor of misinformation"; "He is
not legally regarded as a valuer, nor does he so regard himself"; and
"Laymen with no knowledge of accounting may be deceived or, if they
know the truth, may tend to regard accounting as the weirdest of
professions."

This sat, as you can imagine, poorly with his fellow accountants and
the large accounting firms. Dad paid for his candor.

The definitive article on the cold reception given his proposed re-
forms was written by Stephen A. Zeff, then the outgoing editor of the
Accounting Review, the quarterly journal of the American Accounting
Association, where it appeared in 1982, ten years after Dad died. Its title sums it up: "Truth in Accounting: The Ordeal of Kenneth Mac-
Neal." As I write these words I'm possessed by an adult's awareness of
how deeply hurt he must have been and a youth's memory of how he
bore it in utter silence. Professor Zeff ends with a nice tribute.

Although MacNeal and pioneers before him were seldom appreciated by their peers, they may have given later reformers
the courage that comes from knowing that one is not alone.

Shortly after I came home from Europe in 1946 Dad asked what I
thought of doing with my life. I said I'd like to write. His advice was direct. Some might call it brutal; others, loving. You decide. He said, "I
suggest you wait until you have something to say."

I waited until another father, Father Time, said further delay threatened nonperformance. Mid-life crisis.

A psychoanalyst might suggest this book, and especially this chapter, is my penance for some childhood guilt. I don't think so. But if it
is, well, I hope it works.

On a flight home from the Midwest, I met a man who told me he'd
made his fortune by happening upon a fact nowhere reflected in a
company's accounting statements. The company's sole operating unit
was a distressed department store in a large eastern city. What he had
discovered was that the store's very old, long-term lease contained a
purchase option at a figure far below the current market for the land
alone. With borrowed money he bought the company, exercised the
purchase option, closed the store, sold the property, paid the loan, liq-
uidated the company, and pocketed an enormous profit.

An orthodox Communist would probably have called my seatmate a
leech, someone who adds nothing of social value. In a free enterprise
economy, we regard his activity as useful because through him the
property was put to a more productive use, as evidenced by someone's
valuing it more highly than the store could.

My dad felt it would be easier for everybody to play this socially useful game, and fairer, if financial statements were easier to interpret,
more in line with events, less constrained by the accounting profession's verbal-level conventions.

Elisabeth Ruedy reports that many questions clients ask her as a
math teacher have less to do with math than with commercial conventions. "Neither the law nor the regulations are math," she says. "Ex-
ecuting the regulations and procedures is where the math comes in."

She then adds, "I'm stressing this because knowing it gives you power."

The power Ruedy had in mind was recognizing that, in business, math is only a small part of the numbers, an insight that frees you to ask questions about the conventions used.

Some people use the power. They know there's a gap between an accountant's book figures and market values, a gap that's sometimes filled with gold. They go prospecting.

Real estate accounting is a favorite place to hunt. Dad's first *Truth in Accounting* fable tells one reason why.

THE FABLE OF THE TWO FACTORIES [slightly condensed]. Once upon a time there were two little factories, alike in all respects. Both had been built by a local builder and each was quite obviously worth the same amount, but only the builder knew exactly what it had cost to build them.

In this same locality lived a capable business man named John and a stupid business man named William. The builder of the two factories went to John and succeeded in selling him one factory for $5,000. A few days later this builder went to William and, by reason of William's stupidity, succeeded in selling him the remaining factory for $20,000.

John formed a corporation so he could sell stock to raise money for operating his factory. He sold his factory to this corporation for the same price he had paid for it, namely $5,000 and accepted stock to a par value of $5,000 in payment. John then called in a reputable accountant. The accountant found that John's company had bought a factory for $5,000, and prepared a certified balance sheet showing the factory to be worth $5,000.

William also formed a corporation, to which he sold his factory for the same price he had paid, namely $20,000, for which he accepted stock to a par value of $20,000. William called in the same reputable accountant. The accountant found that William's company had bought a factory for

$20,000, and prepared a certified balance sheet showing the factory to be worth $20,000.

Both John's company and William's company then sold additional stock on the basis of their respective balance sheets. A banker put $5,000 cash into John's company in return for stock to a par value of $5,000, thus acquiring a half interest in John's company for $5,000.

A farmer invested his cash savings of $5,000 in William's company in return for stock to a par value of $5,000, thus acquiring a one-fifth interest in William's company for $5,000.

Now, almost everyone in town, except the farmer, knew that John's factory and William's factory were worth the same amount, so it was not long before William found himself arrested for defrauding the farmer. William defended himself by putting the responsibility for the balance sheet upon the accountant, whereupon the accountant was arrested and put on trial.

The accountant defended himself by confessing that he did not know the value of either factory, nor what they could be sold for, nor what it would cost to build and equip them. He had used the original cost of each factory as its value. He claimed this was the best he could do, a makeshift, but the only makeshift at hand, and challenged the jurors to say what they would have done in his place.

When the jury retired to consider its verdict, it disagreed. Certain jurors thought that the accountant should have made numerous inquiries as to what the factories could be sold for and should have adopted one of the bids as the value of each factory. Some jurors thought that the accountant should have had a builder estimate what it would cost to build each factory, and should have used that amount. Still others felt the accountant did right in using $5,000 for John's factory and $20,000 for William's, because no one could know what the factories could be sold for and an accountant could hardly be expected to know what they could be built for. At the end of three days the jury was still in disagreement and the accountant was released.

But the farmer, nevertheless, because of his reliance on the accountant's balance sheet, received for his $5,000 only a one-fifth interest in William's company, whereas the banker, in reliance on the same accountant's balance sheet, had, for the same sum, received a one-half interest in John's company. Yet William's company had at no time been worth a penny more than John's company although the reputable accountant had certified one as having assets worth $20,000 and the other as having assets worth $5,000.

The accountant was anxious to do right but did not know what to do. Therefore, although he was careful thereafter to keep away from twin factories, he continued to prepare balance sheets in the same manner that he prepared John's and William's. And reputable accountants still do the same down to this day.

Notice how innocent everyone is in this fable. Dad believed that the system, the accounting practices, which confuse symbol-system conventions with market values, were the trouble. Even so, his next two fables illustrated how unscrupulous investors and financiers could use accounting conventions to fleece the unwary.

I won't retell these fables. Nothing my father said in 1939 could possibly compete with the savings and loan debacle of 1989.

You'll recall that in the last chapter I analyzed a reported loss of $500 billion to shoplifting and employee store theft, and—by virtue of its amounting to $2,000 per person, infants and grannies included, and other facts—showed it was impossibly high, perhaps five thousand times too high.

Well, the same figures could well be too low for the savings and loan losses. We, the multitude, have been bled by the few. Some of them, as described by the *Philadelphia Inquirer* of September 9, 1990, "paid themselves lavish salaries and benefits and made loans to their own

dubious ventures and the ventures of friends. . . . Many of the assets—
particularly real estate, shaky loans and junk bonds—are worth far
less than the list price carried on the books of the S&Ls."

In other words, book figures reflect accounting conventions. I
learned that at my father's elbow.

To resolve the savings and loan problems, a four-thousand-person
bureaucracy has been set up with power to hire "outside contractors:
scads of lawyers, accountants, property managers, appraisers, and so
on," many of them, the article notes chillingly, people who "worked
for the savings and loans that failed."

The savings and loan debacle has been called a mess, a financial
mess, a criminal mess, and a political mess. The January 1991 *ABA*
[American Bar Association] *Journal* reports that fourteen of the biggest accounting firms are endangered by savings and loan suits. I call
it a mathsemantic mess. The biggest and worst yet.

Accounting pervades modern life. It affects every business, every
government, and every one of us. Therefore, so do mathsemantic
problems on a grand scale. What was Donald Trump really worth at
the end of 1990? How did Barry Minkow and the ZZZZ Best rug
cleaning empire ever get far enough to be charged with swindling $70
million from banks and investors? "Despite independent audits," reports the *Philadelphia Inquirer* of January 28, 1988, "certifying the au-
thenticity of the company's financial statements, nearly 85 percent of
ZZZZ Best's business was a complete fabrication, said Rep. John
Dingell (D., Mich.), Chairman of the House Energy and Commerce
Committee." I won't moralize about this. You see the point.

Projections and accounting figures reflect a strange and generally
misunderstood fence-sitting mixture of applied and pure math. They
are not true maps of the territories they purport to represent. They
contain omissions and distortions that follow rules relieving their creators of responsibility for accurately estimating actual events. Only in-
siders can use them safely.

I don't condemn planners or accountants. I don't condemn planning
or accounting. I'm a product of both.

I condemn the continued confusion of symbols with events. I condemn the continued substitution of arcane professional symbol-manipulating rules for genuine efforts to estimate events.

Proposition 26: Beware of figures that, like projections and certified financial statements, seem to be meant as estimates of events but which specifically disclaim that responsibility.

CHAPTER

Percentages

I'd like to share with you a practical mathsemantic problem I've had writing this book.

Our recruitment quiz posed five questions on percentages. None was answered correctly by even half the secretarial and clerical applicants. Please recall these applicants had represented themselves as "good at numbers" and passed our initial screening.

The analyst and administrative-law applicants, of course, did better. Most of them had taken some advanced math in school. Even so, their error rate on the five percentage questions varied from more than two out of five to just under one in five.

The questions were in the quiz because I'd noticed troubles with percentages. Nevertheless, before seeing the quiz results I hadn't realized the difficulties were this severe.

Apparently Paulos knew.

> I'm convinced that a sizable minority of adult Americans wouldn't be able to pass a simple test on percentages, decimals, fractions, and conversions from one to another.

Our five questions required converting to percents from ordinary fractions, from decimals, and from a fraction expressed with a division sign. I believe they tested some of what Paulos had in mind. Given our results, my only question about his statement would be whether "minority" is correct.

My writing problem was how to communicate the quiz results. Expressing them in percents, the quiz evidence said, would lose readers.

I considered but rejected the idea of giving an explanation of percentages up front. It would have had to precede the apples-and-oranges example in the first chapter and could well have annoyed both those who needed it and those who didn't.

Patricia Cohen notes an eighteenth-century preference for the form "one out of so many" where today we would use percents. That's exactly what I'd already done for seven chapters before I came across her statement. Call it an unconscious example of mathsemantic cultural perseveration.

If you're a devotee of percentages, you probably noticed my use of the "one out of so many" formula and wondered why I just didn't use percents. Now you know I was scared off by the quiz results. What would you have done in my place?

The five percentage questions in the quiz had the general direction, "Express the following items as percentages to the nearest whole percent."

The best results were obtained with translating common fractions into percents. The first of these was "7/10."

Ninety-five of the one hundred ninety-six clerical and secretarial applicants, not quite half, gave the most frequent answer, which is also the correct one, "70%."

The next most frequent answer, from forty-two applicants, about two in nine, was no answer. I take that to mean they knew they didn't know the answer and didn't want to guess.

That leaves fifty-nine applicants, about three in ten. They gave fourteen other answers, each using one or more of just five symbols, 7, 1, 0, %, and . (decimal point). Of those using a percent sign, two answered "10%"; eighteen, "7%"; ten, "1%"; four ".7%"; one, ".70%";

and one, my favorite, simply " %." Of those forgoing percent signs, five answered "70"; one, "10"; one, "7"; six, "1"; one, "1.0"; five ".7"; three ".70"; and one, ".07."

The five who answered "70" might just have been too lazy to post a percent sign. The other fifty-four, however, were willing to hazard an answer but had only confused ideas of what "7/10" means, or what a percent is, or both. I wish I knew how they got their answers, but I don't.

To anyone who loves words, "percent" explains itself. "Per" means "for each," as in, for example, "one *per* person." "Cent" comes from Latin "centum," a hundred. It shows up in such words as "century," "centennial," and "centenarian," all having to do with a hundred years, "centipede," a hundred-legger, and "centimeter," the hundredth part of a meter.

The best connection for understanding percents, however, is with "cent" as the hundredth part of a dollar. I've used this often in teaching employees. I tell them just to remember that one *cent* is also one *percent* of a dollar. Two cents is two percent of a dollar. Twenty-five cents is twenty-five percent of a dollar, and also a quarter (dollar). Fifty cents is fifty percent of a dollar, and also a half (dollar). A hundred cents is one hundred percent of a dollar.

Then I show employees how to find ten percent, one percent, and a tenth of one percent of any base number. Say the base number is 13,400.

13,400	=	100%, the base number itself
1,340	=	10% of the base
134	=	1% of the base
13.4	=	0.1% of the base

One can now divide any number by 13,400 and express the answer properly as a percent by locating the number's interval in the table. I have employees make a similar table for another base and place a few numbers in the correct intervals. That's usually enough instruction. The rest is practice.

Ninety-five applicants also converted the second common fraction, "3/5," correctly into "60%." Fifty gave no answer. The other fifty-one answers followed more or less the same pattern as for "7/10." The answers most extending the previous range were "3/5%" and "80%."

Converting ".9" gave the next best results. Eighty-four applicants, three in seven, gave the correct answer, "90%." Forty-six, two in nine, gave no answer.

Twenty-two, one in nine, answered "1%," an answer that could show a misunderstanding of decimals, or percentages, or both. Note that the money analogy would treat ".9" as "$.90," ninety cents, ninety percent of a dollar, and therefore as ninety percent.

Eight applicants answered "9%," seven said plain "1," and four said "10%." The answers most extending the previous range were ".009" and "130%."

Converting ".814" gave still poorer results. Sixty-eight applicants, just over one in three, got the correct answer, "81%." Fifty-two, more than one in four, didn't answer. Nineteen answered "80%," showing that rounding got in the way. The oddest answers were "1/8%" and "1000%."

Converting "12 ÷ 3" produced the worst results. Only forty-nine applicants, one in four, gave the correct answer, "400%." Thirty-four didn't answer.

Forty-two said "4%," thereby showing a misunderstanding of division, or percentages, or both. Perhaps applicants didn't recognize "12 ÷ 3" as another way of saying "12/3." I suspect, but can't prove, that some applicants had been drilled in converting fractions to percents, but not fractions expressed this way. If so, they hadn't really understood what they were doing.

I won't describe in detail how the analyst and administrative-law applicants fared. All but five answered all five questions. The questions they answered best followed exactly the same sequence as the secretarial and clerical applicants. Except for the appearance of equivocations, their wrong answers were like those already noted.

Ruedy says percents are "the most important part of practical math," dominating what you need in stores, for personal finances, and on the job. She says one can "learn percents once and for all in five minutes." On the mechanical level, that seems about right.

Her teaching method resembles ours. She explains that "percent" means "per hundred," makes use of the notion that cents are hundredths of dollars, shows how to figure what one and ten percent are, and provides some problems for quick drill.

Paulos says elementary schools teach fractions, decimals, and percentages, but not how to convert from one to another.

Ruedy's book teaches conversions. She begins by stressing that "fractions, percentages, and decimals are different *languages* to express the same numbers." In six pages she then explains all six possible translations (fractions to percentages, fractions to decimals, percentages to decimals, and the reverse of each).

If we think of math as a language, perhaps we should treat fractions, percentages, and decimals as alternative formulations within that language, rather than as separate languages themselves. It's a mathsemantic question worth considering.

Ordinary languages like English are rich in alternative formulations. We use difference voices, such as active ("The dog bit the man") and passive ("The man was bitten by the dog"). We use different moods, such as the indicative ("It's best to start by relaxing") and the imperative ("Please relax").

We translate easily between these formulations. For vigor, stylists advise converting sentences from the passive to the active voice. For clarity and brevity in the office, I phrase instructions in the imperative mood. None of this creates a new language.

If we think of fractions, decimals, and percentages as alternative formulations, we can retain the idea, which I prefer, that they're all part of one mathematical language.

Each formulation within a language presumably serves some purposes better than other formulations do. Otherwise it would disappear. This seems true of different voices and moods and also of fractions, decimals, and percentages.

It's easy to see that "1/2," "0.5," and "50%" all represent "half" and that they translate easily within this little set. Yet the three *forms* are not always so interchangeable.

Take a "third." Its fractional form is 1/3, which conveys the basic relationship of one to three as directly as possible. Everybody who's ever divided a candy bar three ways has literally *felt* the relationship.

The other forms usable for a "third" are less tangible. They create complications. The decimal form is a repeater, 0.33333..., as is the percentage form, 33.333...%. Repeaters require special symbolic conventions. Hand calculators employ inconsistent conventions and don't give simple fractional answers. In a letter to the editor printed in the *New York Times* on December 8, 1990, University of California mathematics professor M. H. Protter advanced these facts as an argument against allowing calculators in Scholastic Aptitude Tests.

Thus the decimal and percentage forms can create symbol-system problems the fractional form avoids. Therefore, fractions will in some circumstances be the preferred form. In other circumstances the preferred form will be decimals and in still others it will be percents.

Math instruction teaches the three forms. It may even show how to translate from one to the other, although many students apparently don't learn how very well.

We also need to ask when each form works best. Why do we usually state some things in fractions and others in decimals or percents? What are the advantages and disadvantages of each form? To what extent is it best to stick with one form and for how long? These questions go beyond math proper. They're mathsemantic questions.

Percents seem to work better than fractions and decimals for expressing comparative rates, changes in rates, and comparative changes in rates.

For example, if you have ten thousand dollars to invest, and can get six percent per year from one investment and seven percent from an equally safe one, then the higher rate will improve your position by one percent, one hundred dollars per year. You could say this in fractions or decimals, but it would be more roundabout, harder to grasp.

The advantage of percentages is even greater when we compare rates having different bases. For example, if a company that was profitable last year has this year increased its income (one base) by 13% and its costs (a separate base) by 9%, then we can say immediately that it's more profitable this year than last. Again, neither fractions nor decimals offer as easy an expression. As a consequence, we use percentages mostly to express rates destined to be compared, especially rates having different bases.

To understand a rate we need to know both what's being counted and the base to which it's being related. A report that housing starts are up three percent in March tells us neither. What are "housing starts"? How are they counted? Are they the actual housing starts in March or counts somehow adjusted for the weather or the season? What does "up three percent" mean? Up from what? Housing starts in February? Housing starts in March of last year?

Many news reports of percentage changes that I hear or read leave me wondering. I'm often unable to judge for myself whether the percent reported is good news or bad. Some reports are clearly wrong; others, just indeterminate.

The *Philadelphia Inquirer* of June 8, 1989, carried a story headlined, "Paramount bid sends Time Inc. stock up 40 pct." The lead ran, "Time Inc.'s stock price yesterday soared more than 40 percent after Paramount Communications Inc.'s $10.7 billion buyout offer." Further on the article stated, "Time yesterday jumped $44 to $170 a share." The opening or base price, then, must have been $126. A $44 increase is about 35%. It looks like somebody mistook the $44 increase for a 44% increase. Of course, one can't be sure. Something else might be wrong.

Aviation Week of March 12, 1990, reported that the international routes of American Airlines "represent about 14% of its total system." I don't know what this means, because I don't know what the base is. Is it route miles, passenger-miles, total passenger and freight revenues, or what? It's my field, I'm an expert, and still I don't know. Perhaps what makes me an expert is knowing that I don't know.

"Of the 64,000 square kilometers of Pamir territory," began a story in the June 1990 issue of *Soviet Life*, "97.5 percent are mountain ridges." That unnerved me. I thought ridges were lines along top edges, crests. A territory consisting almost entirely of edges struck me as a topological impossibility.

Wrong again. The dictionary brought me down nicely. One meaning of "ridge" is "a range of hills or mountains." Thank you, Merriam-Webster.

John Tesh on "Entertainment Tonight" reported that the PBS series "The Civil War" had an audience of 13% versus the usual 4%, "an increase of more than 300%." The percentage-point increase is 13% minus 4%, or 9%. The base is 4%. Therefore the percentage increase is not more than 300%, but only 9/4, or 225%.

Failure to allow for shifting bases leads to bad judgments. If sales drop by 20% and then grow by 25%, you're now better off, right? Wrong. You're back where you started. Proof: $100 minus 20% (one fifth) of $100 is $80. $80 plus 25% (one fourth) of $80 is $100. Add three zeros, or six, to each dollar figure for realism. Same conclusion.

Not all examples are so innocuous. A drop of 70% followed by a rebound of 80% leaves one in deep trouble. Proof: $100 minus 70% is $30. $30 plus 80% is only $54.

A gain of 80% followed by a loss of 70% is equally bad. $100 plus 80% is $180. $180 minus 70% is again $54.

I know. I know. It seems you can't win. That's not true. Many things fluctuate greatly and still gain. But in any given fluctuation, the percent going up has a smaller base than the percent coming down. Therefore, the percent gain must be larger than the percent loss, sometimes much larger, just to stay even.

If this shakes your faith in your ability to judge the effects of percentage changes, consider this: You may be better off for it. Now you know to work out the figures in dollars.

A Civil Aeronautics Board analyst had to recommend which airport, Dodge City or Garden City, Kansas, should be used if the cities consolidated their air service. The two airports offered similar patterns of convenience for their own residents and inconvenience for those of the othcr city. The predominant traffic flows from each city were east to Kansas City, Missouri, and beyond.

The analyst recommended use of Garden City, the more western Kansas airport. This would require most Dodge City passengers to backhaul before starting east. It defied common sense.

The analyst's reason? He'd calculated the local airport inconvenience as a percent of each airport's distance to Kansas City. This percent was lower for Garden City than for Dodge City. Therefore he favored consolidation at Garden City.

What he'd overlooked was that the distance to Kansas City was a variable base, larger from the more distant city. Because the local ground trips were equally inconvenient, dividing by the greater distance necessarily produced a lower percent. The analyst was a good guy. It was an unpleasant duty exposing the mistake.

Some people overlook bases so much they even add percents having independent bases. A. K. Dewdney, writing in the *Scientific American* of November 1990, reports a particularly flagrant example. A survey had shown that 49% of the Italian male and 21% of the Italian female respondents confessed to extramarital affairs. This led to the newspaper headline, " SEVEN ITALIANS OUT OF TEN HAVE COMMITTED ADULTERY. "

Not true. The percent for a combined base must lie *between* the percents for the separate bases. Thus we don't add 49% and 21%, but, assuming equal numbers of males and females, we take the average, 35%.

If the bases aren't equal, the combined percent will be closer to that with the larger base. Overlooking this leads to the famous horse-and-rabbit stew error. A traveler was invited to have some horse-and-rabbit stew. "What," he inquired, "are the proportions?" "Fifty-fifty," he was told, "one horse, one rabbit."

The eleven largest air carriers (the majors) and the next nine (the nationals) showed a combined traffic growth in July 1990 of 7.4% from the previous July. The traffic of the nine nationals alone increased by 23%. "Much of the growth," concluded *Aviation Daily*, "was fueled by traffic surges among the national carriers."

Not so. It's a horse-and-rabbit stew. The eleven major carriers combined carried about forty times the traffic of the nine nationals. The majors hadn't grown at as fast a rate, true, but they still contributed more than nine-tenths of the traffic increase.

The growth *rates* of some small companies typically exceed that of all larger companies, while the *absolute* growth of most larger companies exceeds that of all smaller companies.

Company Size	Base Sales	Absolute Sales Growth	Sales Growth Rate
Small	$100,000	$50,000	50%
Large	$10,000,000,000	$100,000,000	1%

Higher growth rates don't even necessarily mean that the smaller companies are catching up. In the example given above, the smaller company has just fallen further behind by $99,950,000.

Comprehension of rates, of one thing divided by another, is a considerable mathsemantic advance over comprehension of simple quantities. Recall the difficulty Piaget said children have understanding *speed*. They visualize it as the act of overtaking rather than as miles divided by time.

Conceptualizing comparative rates, changes in rates, and comparisons of changes in rates must be even more difficult than conceptualizing rates. Few people, apparently, acquire the knack. Errors abound.

Please plow through the following example.

The *Aviation Daily* of December 13, 1990, reported that "Eastern's November 1990 traffic declined 5.3% from a year ago on 6.9% fewer available seat miles, causing the airline's load factor to decline one percentage point." This is impossible.

"Traffic" here means passenger-miles. (You'll recall the difficulty applicants had with multiplying unlike things. So we're already off to a tough start.) Passenger-miles decreased by 5.3%. Available seat-miles (which I hyphenate) decreased by 6.9%. Therefore, passenger-miles decreased proportionately *less* than seat-miles.

"Load factor" means the percent that passenger-miles are of seat-miles. If all seats were always filled, the load factor would be one hundred percent. If filled half the time, the load factor would be fifty percent.

So, if passenger-miles decreased proportionately less than seat-miles, then the load factor would, contrary to the story, go up, not down. Indeed, an accompanying table showed this is what actually happened.

If you can follow what I've just said, congratulations. If not, don't worry. The point I'm trying to make is that understanding how percents are used is not easy. Practically nobody receives the necessary training.

You may recall my mentioning that Ruedy said one could learn percents in five minutes and my agreeing that was about right for the *mechanical* level. As you can see, the pure math of percents may be easy, but their mathsemantic uses are hard.

Percents in the abstract, in the pure math sense, are trivial, merely a change in the form of expressing a division. Five minutes is enough to learn them. Understanding the ways percents are used is definitely not trivial for living in today's world. Five minutes is definitely not enough. Practice with real percents is needed.

But not just with baseball statistics. That's not good enough.

Philadelphians say that Rich Ashburn (also known as Whitey, and as Richie during his playing days) should be elected to the Hall of Fame. Among other things, he was the batting champion in 1955, hitting three thirty-eight, and again in 1958, when he hit three fifty. If you look these figures up, perhaps in the *World Almanac*, you'll find them in a column unambiguously headed "Pct.," but then reported as ".338" and ".350" in decimal form. On occasion someone will say, "One year Ashburn hit three hundred and fifty percent." Just think. Rich Ashburn hit three hundred and fifty percent! and he's not in Baseball's Hall of Fame.

Perhaps what we need is a name for "per thousand." Baseball isn't the only industry using "per thousands." Direct mail, for example, quotes mailing costs that way. What word could we use? Well, just as a "cent" is a hundredth part of a dollar, a "mill" is a thousandth part.

A millipede is a thousand-legger. So how about "permillage"? It's in the dictionary. "Rich Ashburn hit three hundred and fifty permill." On second thought, forget it.

I suspect some reporters, perhaps partly misled by baseball statistics, lack the mathsemantic sophistication necessary to judge the percents appearing in their stories. Many otherwise sharp copy editors seem to have the same difficulty. That leaves them and us more at the mercy of whoever issues the figures.

Dewdney shows just how great the risk may be.

> Once one puts it all together, it is easy to see how the media can play up the numbers, an all-too-common abuse I call numerical inflation. Joseph Childers of Bryte, Calif., has passed along documented claims about the percentages of fatal traffic accidents having various causes: cocaine, 20 percent (a New York newspaper); marijuana, 25 percent (Drug Enforcement Administration); alcohol, 50 percent (California Highway Patrol); sleepiness, 35 percent (sleep researchers); speeding, 85 percent (National Transportation Safety Board); smoking, 50 percent (National Highway Safety Administration); suicide, 35 percent (suicide researchers); mechanical failure, 20 percent (New York State Department of Motor Vehicles).

The total is 320%. That seems to leave no room for bad judgment in driving accidents but plenty in our thinking about them.

The same kind of room seems available in our thinking about crime, prices, the environment, international finance, company performance, population growth, marketing claims, war, and just about everything making the headlines these days.

> **Proposition 27:** Percentages are dangerous social and economic tools that appear easy only to math teachers and the inexperienced.

CHAPTER

Surveys

H ere's a question for you. Who's the most famous, the best known statistician the world has ever seen? Take your time. No hurry. Oh, and aloud, please.

If you mentioned *any* of these, you pass gloriously: John Graunt, Jacob (aka Jakob, James, Jacques) Bernoulli, Edmund Halley, Abraham De Moivre, Thomas Bayes, Thomas Simpson, Roger Joseph Boscovich (aka Rudjer J. Bŏsković), Tobias Mayer, Pierre Simon Laplace, Adrien Marie Legendre, Carl Friedrich Gauss, Siméon Denis Poisson, Adolphe Quetelet, Gustav Theodor Fechner, Antoine-Augustin Cournot, Francis Galton, Wilhelm Lexis, Francis Ysidro Edgeworth, Hermann Ebbinghaus, Karl Pearson, George Udny Yule, Wesley Clair Mitchell, Ronald A. Fisher, Frederick C. Mills, George W. Snedecor, Jerzy Neyman, Frank Yates, W. Edwards Deming, M. S. Bartlett, R. G. D. Allen, Abraham Wald, Arthur F. Burns, Maurice G. Kendall, John W. Tukey, Harold Cramér, William G. Cochran, Janet Norwood, Barbara A. Bailar.

If you couldn't name a single one, don't let it upset you. Even the article on statistics in the *Columbia Encyclopedia* doesn't name a statistician.

The answer, however, is none of these. The honor of being the best known, not necessarily the best, statistician falls to Florence Nightingale. Yes, *that* F. N.

Here's just part of what the *American Statistician* of May 1990 said about her:

> Florence Nightingale, the "passionate" statistician, was born May 12, 1820, in Florence, Italy (after which she is named) to upperclass English parents and died August 13, 1910. She received excellent early preparation in mathematics from her father and favorite aunt. She was tutored by mathematician James Sylvester (who developed the theory of invariants with Arthur Cayley), and she was reputed to be his most distinguished pupil. She was also influenced by the work of Adolphe Quetelet, a Belgian scientist who applied statistical methods to data from several fields, including "moral" statistics (social sciences).
>
> In response to a "call from God," Nightingale decided to pursue nursing as a career.... In 1854 she was named by Lord Sidney Herbert (England's secretary at war) to be a nursing administrator ("Superintendent of the Female Nursing Establishment of the English General Hospitals in Turkey") for a group of 40 nurses in Scutari during the Crimean War. No woman had ever been so distinguished before.
>
> Nightingale was an innovator in the collection, tabulation, interpretation, and graphical display of descriptive social statistics. Her "Notes on Matters Affecting the Health, Efficiency, and Hospital Administration of the British Army" showed that, among other things, the number of soldiers killed by disease was more than seven times greater than those felled in battle. She made an urgent plea for keeping standardized medical statistics and advocated establishment of a Statistical Department of the Army. Her reports were regularly read by Queen Victoria, who awarded her the St. George's Cross.

She was a pioneer in the graphic illustration of statistics. . . .

She became a Fellow of the Royal Statistical Society in 1858. She advocated recording health and housing information in the 1861 census. She introduced . . . standard data-collection forms, which were used by hospitals to record vital statistics, and she drew up a program . . . for the 1860 International Statistical Congress . . . published in the *Journal of the Royal Statistical Society* in 1862.

She became an honorary member of ASA [the American Statistical Association] in 1874. . . .

Nightingale zealously pursued establishment of an Oxford professorship in applied statistics. This would have been the first university teaching of statistics ("social physics"). There was much correspondence between Francis Galton and her on the subject. For a variety of reasons (financial, and the fact that the subject was not covered in university examinations) the project was eventually abandoned.

Karl Pearson acknowledged her as a "prophetess" in the development of applied statistics. In 1857, the year of Pearson's birth, she was already using tables of observed and expected mortality counts for British soldiers.

Statistics was, she proclaimed,

> the most important science in the whole world, for upon it depends the practical application of every other (science) and of every art: the one science essential to all political and social administration, all education, for it only gives exact results of our experience. . . . To understand God's thoughts, we must study statistics, for these are the measure of his purpose.

The study of statistics was, for her, a religious duty.

Our remembrance of Florence Nightingale as a nurse rather than as an administrator or statistician reflects the semantic habit of stereotyping. We could ask, "Had Nightingale's achievements been accomplished by a man, what is the probability he would now be remembered as a male nurse?" Vanishingly small, wouldn't you say?

Math-statistical prominence may have been unusual for women a hundred years ago. But not today. In 1990 the executive director of the American Statistical Association was mathematical statistician Barbara A. Bailar, and the Commissioner of the U.S. Bureau of Labor Statistics was Janet L. Norwood. Other women with distinguished math-statistical careers in the United States have included, among others, Gertrude M. Cox (Fellow of ASA, the Institute of Mathematical Statistics, etc.), Florence N. David, (aka F. N. David, author of seven books on statistics and more than a hundred scientific journal papers), Margaret E. Martin (first director of the Committee on National Statistics of the National Research Council), Elizabeth L. Scott, (chair, Statistics Department, University of California, 1968–73, among other posts), Helen Mary Walker (author, teacher, and in 1944 first woman president of ASA), and Arness Joy Wickens (distinguished career with National Emergency Council, U.S. Central Statistics Board, U.S. Department of Labor, United Nations Economic and Social Council, and in 1952 president of ASA).

If you didn't realize that women could excel at math, you've been a victim of semantic stereotyping.

Just as we at times expect too little, so we can at others expect too much.

Lorence E. Moore was one of the best sales-group heads the direct-mail firm of O. E. McIntyre, Inc., ever had. Some of his clients were the Bankers Life & Casualty Co., where the MacArthur Foundation "genius" award money was made, *TV Guide*, and *Reader's Digest*, the circulation leader of the time. Part of the *Digest*'s success was owed to the brilliant campaigns run by Larry, and perhaps the most brilliant was the penny mailing.

The idea of the penny mailing—and so far as I know it was a mass-mailing "first"—was to have two shiny coins beckon through a window, thereby almost insuring that the recipient would open the envelope. Avoiding the peremptory toss into the circular file is the direct-mail advertiser's first aim.

The first test, if memory serves me correctly, was a million pieces. This required two million pennies. The Mint refused to supply them. So Larry set up a collection system paying a small premium to tradespeople for their pennies. A shortage of pennies on Long Island ensued. But Larry got what he needed. The pennies were then washed in preparation to being hand-tipped onto the mailing. So far, so good.

The department chief handling the penny tipping called Larry to find out if he wanted all the pennies face up, which would have looked especially nice. Practical Larry replied, "No, half up and half down is okay." He was perfectly content with a chance distribution of approximately fifty-fifty, heads-tails.

The next thing anybody knew, the job was falling further and further behind schedule. After several futile calls, Larry went to the plant to observe directly. There he found the trouble. The penny tippers were improving on chance. They were guaranteeing a half-and-half distribution. On each mailing piece, they were carefully placing one penny heads up and one penny tails up.

My experience as a direct-mail research director had alerted me to people's troubles with chance, probability, sampling, surveys, statistics, and the like. So I included three statistical questions in the version of the recruitment quiz given to the one hundred ninety-nine applicants for analytical and legal jobs. You don't have to understand statistics to get the morals I'm going to draw from them.

First question:

1. Which is generally the most important in determining the reliability of a sample: ( ) the absolute size of the sample, or ( ) the percentage which the sample is of the entire population, or ( ) the absolute size of the entire population.

Five applicants, one in forty, chose the third option, the one that makes the least sense. One doesn't need statistical training to see that the size of a population generally tells you nothing about the reliability of an otherwise unspecified sample.

One hundred seventy applicants, six of seven, chose the percentage option. Unfortunately, it's the wrong answer. Knowing that a sample is, for example, two percent of the population tells you virtually nothing about its reliability. A two percent sample of fifty students would be one student, obviously unreliable. A two percent sample of United States adults would be about 3,500,000 adults, far more reliable than needed for most purposes.

If the percentage were critical, we'd generally be out of luck. The usual assumption made in statistical inference is that samples are drawn from infinite universes. There's no way for a sample to be *any* percent of an infinite universe.

I sometimes asked applicants who'd given the percentage answer, "How would you estimate the reliability of a sample of coin flips?" "Well," they'd answer, "a two percent sample wouldn't be as good as a three percent sample." "But," I'd inquire, "two or three percent of what?" "Of the total coin flips," they'd say. "Oh," I'd say in mock wonder. "How do we find out how many that is?" Nobody ever said.

If they'd studied statistics, I might ask them for a formula relating to reliability, say, the standard deviation. Now, it so happens that all the easy formulas contain terms for the sample but not for the universe. So when we'd agreed on a formula, they'd then see that it contained no term for the universe, yet it measured reliability.

Only twenty-four applicants, one in eight, chose the absolute-sample-size option, the right one. It's easily shown that a sample's absolute size must beat its percentage size in measuring reliability. A *two* percent sample of a universe of fifty is one. A *one* percent sample of a universe of a million is ten thousand. A sample of ten thousand beats a sample of one. End of story.

What's amazing is that, after eliminating the five applicants who chose option three, if the rest had simply flipped a coin, about ninety-seven would have answered correctly, not just twenty-four. What produces this bias for the wrong answer? Could it stem from hearing such phrases as "a two percent sample"? It seems too powerful for that. Perhaps bad mathsemantics allows "two percent" to refer to objects rather than to a rate. In any event, statistics courses apparently don't cure it.

Other things being equal, small samples are less reliable than large samples. This has unappreciated consequences. For example, Berlin, Begg, and Louis in the June 1989 *Journal of the American Statistical Association (JASA)* look at published reports of cancer clinical trials and find "small studies appearing to possess large treatment effects and large studies possessing relatively small effects."

Apparently what happens is this: Many small trials are run on potential cancer therapies. Random statistical error helps make some therapies look better than others. The small trials seeming to show the most beneficial effects are, understandably, the most likely to be written up and published. Larger trials are then run and usually show fewer beneficial effects than the smaller trials.

Thus, most advance reports based on small trials will not be confirmed by larger trials and should be interpreted with skepticism.

Second question:

2. How good a fit does a coefficient of correlation of 1.0 indicate?_____ _____

If you knew nothing at all about statistical correlation, it's unlikely you'd get this right. However, one hundred and four applicants, more than half, *did* get it right. They said in various ways that it represented a "perfect" fit.

Another forty-seven, about one in four, said it was a good to excellent fit. Thirteen, about one in fifteen, said it was a fair to poor fit. Thirty-five, about one in six, didn't answer.

Now, how could more than half the applicants know something as esoteric as what a coefficient of correlation of 1.0 means, yet only one in eight realize that the absolute size of a sample determines its reliability more than its percentage size? The explanation, I believe, is similar to that for the algebra problems. The coefficient of correlation was learned as pure math. It had less chance than the reliability question to get tangled in mathsemantic misunderstanding.

Third question:

3. In general practice, which is the most important to the reliability of a sample: _____ size of sample; _____ control of sampling bias; _____ survey method (i.e., mail, telephone, personal interview, etc.); _____ quality of editing; or _____ tabulating method.

The most frequent answer, selected by one hundred twenty-two applicants, just over three out of five, was the one hoped for, "control of sampling bias."

If you don't control bias, it doesn't matter how big the sample is. The famous *Literary Digest* poll of 1936 covered more than two million respondents, yet because of failure to control wealthier-respondent-self-selection bias, picked Landon over Roosevelt.

Forty applicants, one in five, picked "size of sample." They apparently hadn't heard of the *Literary Digest* case and couldn't figure the answer out for themselves.

Twenty-two, one in nine, picked "survey method." They might have asked themselves which they'd rather have, a biased sample queried by the best method or an unbiased sample queried by another method.

Five picked "tabulating method," five didn't answer, four split their answers, and one applicant picked "quality of editing." "Secretly," said one of my editors, "I'm touched."

I think one reason applicants did better with this third question is that it stayed on one side, the semantics side, of the mathsemantics divide. The question could be answered almost without considering math.

On the three statistical questions, applicants performed best (122 right) when *no math* was required, next best (104 right) when *only math* was required, and by far the worst (only 24 right) when *both math and semantics* were required. Three questions make a tiny sample, of course, but the results suggest once again that we are ill-prepared to handle problems involving both math and semantics.

It's a mistake to expect one perfect survey to act as a crucial experiment, to provide absolutely conclusive findings. There are always objections that can be made by interested persons on scientific grounds about samples, data collection, and analysis. When enough studies go the same way, however, as in the findings about the hazards of smoking, statisticians speak of "robustness." A series of individually inconclusive surveys can, taken together, lead to robust conclusions.

Surveys proposed to prove generalities once and for all tend to be huge and unworkable. They can easily become expensive examples of the single-instance habit operating in unbounded domains. A series of small surveys filling in the voids in a knowledge web works much better.

In our aviation work, for example, we start our estimates from the nearest strand on the web. If we want to estimate how many passengers a new airline service will generate, we start with current passenger counts in the city-pair market, even if only connecting-plane service is offered, and work from there. We've made such estimates so many times, that now we seldom need a survey.

Sometimes, however, new services involve quantum jumps to novel conditions. This happened when Midway Airlines started services competing with those at O'Hare. It happened when People Express introduced sharply lower fares on services to the New York area via Newark. In these cases, a survey helped.

Nevertheless, when we survey, we stick as close to our knowledge web as we can. The last thing we'd do would be to start asking a properly random sample of all people whether they'd use the new service. That would be inordinately expensive and it might also be quite misleading. Instead, we survey people who should be interested—who already travel to New York, for example—and we provide an extensional context enabling them to work within their knowledge webs.

We ask, for example, "What was the date of your *last* trip to New York?" That established, we ask, " *Since* that trip, have you had any occasion to postpone or avoid a New York trip that you would have taken

had the following service and fares [which we then specify] been available?" We provide boxes to check off, "Yes, definitely," "Yes, probably," "No, probably," and "No, definitely." If "Yes, definitely" is
checked, we then ask, "How many times?"

These data enable us to figure for these respondents the ratio of
their expected new trips to their existing trips if they were offered new
service and fares. That ratio is *all* we want from the survey. Everything
else is already in our web. We apply the ratio to the known number of
existing trips to make a reasonable estimate for the entire city-pair
market.

Such a survey forces respondents to work within their own *knowledge* webs. They may ardently desire the proposed improvements.
However, if they've just returned from a New York trip, our survey
obliges them to say they've not postponed or avoided any trips they
would have taken with the improvements. A few so object to having
their desire thus masked that they add a note telling us what a dumb
survey we've designed. We treasure the notes as proving the survey
worked.

Professor A. S. C. Ehrenberg, writing in the *American Statistician* of
August 1990, states that statisticians need to pay less attention to the
single study and more attention to the analysis of many sets of relevant
data. I agree. The single study engages in a once-and-for-all struggle to
establish the significance of some new fact. The analysis of many sets
of data first sets up new strands in one's estimating web and then ad-
justs them. It's cheaper, faster, more flexible, and less prone to a once-
and-for-all error.

A common example is the election-day exit poll. The particular poll
means little in itself. However, comparing it with previous election-day
exit polls shows what shifts have taken place. These shifts can then
be used to adjust the knowledge web of voting behavior to produce a
forecast.

Paulos, who admits to only informal polling experience, seems to
overlook the use of surveys to adjust knowledge webs. His criticisms
apply mostly to once-and-for-all studies. For example, he states, "Unless your poll sample is randomly selected ... results usually mean

very little." This works as a criticism of the biased respondent selection in the 1936 *Literary Digest* poll considered as a once-and-for-all survey. No argument there.

However, had the survey been used to adjust known voting results, as exit polls do today, it could have shown that the respondents intended to elect Landon in 1936 by a *smaller* margin than they'd intended to elect Hoover in 1932, presaging a *bigger* landslide for Roosevelt in 1936 than in 1932.

Therefore, the survey's main fault was its once-and-for-all design that required but then did not obtain a random sample of all voters. A much smaller survey intended only to adjust a knowledge web would have cost less and given a correct result. Correcting for distortion is usually much cheaper than trying to eliminate it altogether.

I believe many statisticians know this. Perhaps one reason we don't hear more about it, other than for exit polls, is that a big survey commands more attention than a small survey. I don't know. I've been too busy looking the other way, trying to do more with less.

In the direct-mail business, Larry Moore once asked me, as research director, to design a long series of monthly mailings to determine the seasonal pattern of percentage return to *Reader's Digest* subscription offers. To pique my interest, he showed me the results of more than half a dozen earlier efforts that had failed. In each case there was some problem that had invalidated the results. Some efforts hadn't controlled the monthly mailings for day of week, some hadn't controlled for holidays, some hadn't used the same mailing piece throughout, some hadn't used addresses of equal freshness, some hadn't distributed the names properly into the monthly mailings, and some hadn't even managed to mail every month. The efforts also differed in length, from six to eighteen months.

I knew Larry expected me to design a U.S. survey to run perhaps two years, with a subsurvey for Canada. The total mailing quantity would be several million pieces.

I took the existing data and analyzed them statistically. Although each series was flawed, with a little savvy and some standard statistical techniques, I was able to derive the average seasonal variation in sub-

scription return. Canada's resembled that for the United States, except
with greater swings between the best mailing months (January, for example) and the worst (June, for example).

The analysis was so convincing that Larry couldn't justify doing
another series of research mailings. He used my report to build character with *Reader's Digest*. But he lost a big order that was as good—
except for a frugal research director—as in the bag.

I've performed no once-and-for-all studies. Instead, I've done repeated modest studies. I haven't used any extraordinary statistical
techniques, just such ordinary ones as time series and multiple
correlation. Basically I've concentrated on getting the mathsemantics
right.

Such studies have little current academic interest. Mine are neither
quoted nor challenged. They don't fit what has been called "the cult of
the single study." They make mathsemantic demands. Yet I'm informed that my aviation book has been seen on the desks of academics
who theorize about the aviation matters I've studied. I'm still hoping
that studies based on multiple data bases and sound mathsemantics
will come into vogue. Mine may yet merit a footnote.

I'm not griping. I've been amply paid by clients for my studies. My
career is mostly behind me. I just wonder if academics aren't overlooking something.

In addition to the problems surveys have with sampling and web-
bing, there are the problems that occur during data collection, editing,
and analysis. Errors of every conceivable kind do occur.

The 1840 U.S. census produced the finding that free northern
blacks had a much higher insanity rate (1 in 162) than southern slaves
(1 in 1,558). This was interpreted as showing the benevolence of slav-
ery. Apparently the complex eighty-column form used in the field led
repeatedly to posting of small numbers of "idiot whites" in the poorly
labeled "idiot colored" column. This error had a smaller effect on the
southern statistics than on the northern ones, because the colored
population, the base for the ratio, was much larger in the south. The
error went unnoticed. Central editing in that era was confined to

checking that totals had been added properly. The figures were published as facts.

Cohen calls the 1840 census, or at least the controversy it stimulated, "a landmark event" in American numeracy. "Skepticism about the reliability of numbers replaced the earlier naive view that all numerical facts were sacrosanct."

Editing has come far since 1840. I recall doing a gasoline service-station survey that seemed to yield odd results. Internal checks narrowed our doubts to the data obtained during one shift at one station. We questioned the interviewing firm. The next day we received an abject apology and were advised that contrary to instructions, the trained interviewer for that station-shift, in order to handle some personal business, had illegitimately subcontracted his duties to a friend.

Analysis, however, opens Pandora's box.

Out may come a too trusting analyst, perhaps to work on the Gallup Organization's big annual random survey for the Air Transport Association. The 1990 report earnestly states that the percentage of American adults who have *ever* flown dropped from 78% in 1989 to 74% in 1990. The trouble is the drop happens to exceed the adult population turnover rate.

Out also come untrusting analysts who frame surveys so that partial reportings of results create misleading conclusions. Dewdney describes a survey of doctors regarding three headache remedies, brands X, Y, and Z, the last a close variant of Y. Brands Y and Z combined were preferred by the doctors to X. But their preference for brand X was greater than that for brand Y alone, and that is what the manufacturer reported.

Users of some brands can't tell them from others without seeing the label. Dewdney tells how this can be used in blindfold tests to "prove" that half of those who drink brand X prefer it to their own brand.

Becker, Denby, McGill, and Wilks, in "Analysis of Data from the *Places Rated Almanac*," from the August 1987 issue of *American Statistician*, give an example of how far analysis can go. The 1985 *Places Rated Almanac* had rated three hundred twenty-nine places on the basis of nine variables, namely, climate mildness, housing costs, health care and environment, crime, transportation supply, educational opportunities and effort, arts and cultural facilities, recreational opportunities, and personal economic outlook.

Each of the nine variables is itself a composite of various factors. For example, the housing-costs variable is the sum of three components: utility bills, property taxes, and mortgage payments. It can be attacked for excluding apartments and for other failings. The other variables are even more questionable. Nevertheless, the composition of the variables isn't the most interesting point.

The most interesting point is the analysis of the answers to the questions we are quickest to ask: "How do the cities rank? Which city is best? How did my city do?" The answers depend, of course, on how the nine variables are combined. In technical terms, this is a function of the weights assigned to each variable to produce the final ranking. The choice of weights, however, is arbitrary. Some people value a mild climate above all else; others, low housing costs; and so on.

The article ran in a statistical publication; hence the authors make a statistical point. They show that, *depending on the choice of weights*, any of one hundred thirty-four cities could be ranked as number one and any of one hundred fifty could be ranked dead last, number three hundred twenty-nine. Indeed, depending on the weights, fifty-nine cities could have been ranked *either* first or last.

New York, for example, would have ranked first with the following percentage distribution of weights (which do not add to 100% because of rounding): Health—24%, Transportation—22%, Arts—22%, Recreation—17%, Housing—14%, and Climate—2%. New York would have ranked last with these weights: Crime—36%, Housing—34%, Economics—18%, Climate—6%, and Education—5%.

However, the following weights would have caused Los Angeles to rank first: Health—24%, Arts—22%, Climate—21%, Housing—17%, and Recreation—16%. Los Angeles would have ranked last with these weights: Housing—34%, Crime—34%, Economics—18%, and Transportation—14%.

Thus New York does better than Los Angeles when more weight is given to transportation and worse when more weight is given to climate. What's new?

Here are some weights needed for various metropolitan areas to rank first: Oakland, Calif., Climate—82%; Anderson, S.C., Housing—82%; Detroit, Mich., Health—81%; Wheeling, W.Va., Crime—79%; and Fort Collins, Colo., Recreation—82%.

To rank last would take these weights: Bismarck, N.D., Climate—60%; Santa Barbara, Calif., Housing—78%; St. Joseph, Mo., Health—69%; Miami, Fla., Crime—96%; Nassau, N.Y., Transportation—74%; Lafayette, La., Education—50%; Fort Walton Beach, Fla., Arts—86%; Fayetteville, N.C., Recreation—82%; and Jackson, Mich., Economics—66%.

A troublesome mathsemantic concept central to surveys and statistics is that of uncertainty, or chance. The quantification of uncertainty is called "probability." We express it mathematically on a scale from zero (meaning absolutely no likelihood) to one (meaning certainty) or 100% (meaning the same thing). We express an equal probability as one chance in two, 1/2, a 50% chance, 50/50, or .5. The form of expression is arbitrary. All mean the same thing.

But there are deeper difficulties with probability. The Bible says neither the swift, the strong, the wise, the understanding, nor those of skill escape, "but time and chance happeneth to them all." Voltaire countered, "Chance is a word void of sense; nothing can exist without a cause."

Chance also has no chance with predestinarians, those for whom everything happens by plan. What's the use of mortality statistics if each being dies, as seventeenth-century Puritans believed, at its appointed hour?

Much of what Paulos says in *Innumeracy* concerns our inability to handle probabilities. He cites zealots and fanatics who, believing some reason must lie behind every coincidence, "seldom hold any truck with anything as wishy-washy as probability." He gives many engaging examples of how our ordinary expectations differ from actual probabilities and mislead our thinking about events. He notes that "an appreciation for probability takes a long time to develop" but is invaluable for a better command of known facts.

I direct the reader to the beliefs of children discussed in chapter 4. Piaget discovered that for children there is a reason for everything, nothing happens by chance.

We have earlier seen how the statement, "You can't add apples and oranges," coupled with the child's belief in names as essences, disrupts our abilities in addition. Perhaps the oft-repeated statement, "There's a reason for everything," coupled with the child's belief that nothing happens by chance blocks our understanding of probability.

I bring this up not to resolve the questions surrounding probability, but to point out that it's a central mathsemantic problem. It affects not just individuals but the whole nation.

Perhaps the surest sign that we are becoming a society run by numbers is how statistics now affect our legal disputes. What is a life worth? What is the chance that exposure to a chemical many years ago caused a disability? Does a statistical imbalance in employment by ethnic group constitute unlawful discrimination? Do the data prove the defendant is the actual father? Does a particular series of stock transactions prove fraud? How can figures be used to greatest advantage in court? Is statistics, as the promotion for the book *Statistics for Lawyers* claims, now at the cutting edge of legal problems?

The importance of statistics in our lives continues to grow. We will need far more people, at least those graced like our analyst applicants with advanced courses in math, able to handle statistical questions with mathsemantic common sense.

> **Proposition 28:** Understanding statistics is difficult yet vital; difficult because statistics bear so many mathsemantic burdens, and vital because statistics increasingly guide our lives.

CHAPTER

The bugs in the bottle

Except for my faith in human ingenuity, I'd probably be rather discouraged.

Look at some of the challenges we face. Global warming, destruction of the protective ozone layer, ocean pollution, destruction of the tropical rain forest, air pollution, extinction of animal and plant species, groundwater pollution, urban crowding, exhaustion of natural resources, desertification, drug-resistant disease organisms, aging populations, expanding slums, widespread poverty, global economic imbalance, increasing popular expectations, increasing crime, increasing demands for energy and other resources, and ever-faster absolute population growth.

Our choices, individual and collective, deliberate and habitual, totaling trillions a day, carry us relentlessly on. We can never revisit the *status quo ante*.

Understanding such challenges requires the ability to comprehend large numbers, to know what they count or measure; to understand rates, rates of change, and accelerated rates of change; to estimate; to gauge time; to forecast, and to preview how the changes may affect our lives. In short, just to grasp the nature of the challenges we face requires considerable mathsemantic sophistication.

Unfortunately, such sophistication is not biologically innate. Our species evolved handling small-scale, short-term, life-and-death situations. Mathsemantics played no significant role. Our progenitors had no need to comprehend the ultimate cumulative effects of very large numbers of small events, the repeated ventings of chlorofluorocarbons, the repeated junkings of plastic packaging. The role we billions play as the major influence on earth's overall destiny is new. Criticisms have only just started to come in. There hasn't been enough time or feedback for biological evolution to build the needed lessons into us.

Nor does mathsemantic sophistication reach us at our mother's knee through the magic of cultural evolution. Our language suggests we can't add apples and oranges. It doesn't warn us against the difference between repeated actions and single instances. It doesn't require us to see tomorrows as continuations, but permits us to consider every tomorrow as a new beginning.

Our culture, when it regards math as a language, treats it as unique, exalted, precise, difficult. Those who speak math well are regarded with awe while they speak it. But our culture divorces math from the questions regarding meaning that we have learned to ask of our other languages.

Even many people who consider themselves "good at numbers" won't add things having different names, don't really know what they're counting, can't handle zero, don't sense repeated events as different from the people performing them, count such events as people, have virtually no feel for large numbers, can't combine dimensions to reflect interactions, don't understand maximizing well enough even to take costless early action, won't estimate, don't recognize how the characteristics of symbol systems differ from events, take accounting conventions at face value, have little feel for rates of change, don't really "get" percentages, fear to challenge numbers openly, don't grasp the quantification of uncertainty, don't look behind statistics, misjudge the cumulative effects of repetitive events by using single-instance stereotypes, and exhibit other difficulties stemming from a divorce of numbers from meanings.

Higher education is no safe cure. Many who've taken advanced math classes still exhibit mathsemantic difficulties.

Indeed, the mathsemantic ability of our species is so underdeveloped that we should probably regard our ever having-to-rely-on-an-appreciation-of-numbers-for-survival as a sign we're in serious trouble.

If mathsemantic sophistication isn't instilled by biology, fostered by unconscious culture, or required by formal education, we may well ask, "Where does it come from?" I don't know. So far as I can tell, it arises from chance, unusual parents and friends, odd interests, a rare teacher, independent thinking, or all five.

Most people don't know there is such a thing as mathsemantic sophistication. Many applicants expressed surprise at our recruitment quiz. "Weird," they said. "Dumb." "Stupid." "Unfair." "Really *unfair.*" "I've never seen anything like it." "Where'd you ever get such crazy questions?" "You must be wrong; otherwise I'd have heard of these things before." One cut was particularly elegant. I adopted it for use in greeting applicants who'd done well enough to get to my office. "How'd you like," I'd say smilingly, "our idiosyncratic quiz?"

I don't think we dare wait for institutionalized education to evolve enough to teach us mathsemantic sophistication. In the normal course of events, we'd need to allow perhaps twenty years for educators to agree on the importance of mathsemantics, perhaps another ten to decide whether it should be taught by math teachers or English teachers, another twenty to train sufficient teachers, another twenty for them to train the first generation of mathsemantically sophisticated future leaders, and then perhaps another twenty-five for that full generation to reach actual leadership. Thus, after ninety-five years we could begin. That really won't do, will it?

If Kamii is right, children's math would blossom in first grade with an open curriculum of games, but schools are geared to test scores and orderly, quiet classrooms. Thus, we're not allowing our kids today the intellectual autonomy they need to develop a good feel even for basic arithmetic.

If my ruminations stopped here, yes, I'd be discouraged.

But then there's good old human ingenuity. And a lot of other wonderful traits fostered by biological and cultural evolution.

I believe that once enough of us really see mathsemantic sophistication as a beneficial and acquirable cognitive skill, things will move fast.

The Cherokee farmer-silversmith, Sequoyah, also known as George Guess, could not speak English or read or write any language, yet the "talking leaves" of the white man inspired him to devise an eighty-five-character writing system for his people. For twelve years he suffered ridicule. His wife, after taking ten years of that, burned all his papers.

Nevertheless, two years later, in 1821, he was ready for a demonstration. Sequoyah went to one end of the village; his young daughter, Akoya, to the other. Villagers dictated messages, which father and daughter scratched on bark. Runners carried the messages from one to the other. Each then successfully read what the other had written. A few repeats convinced the onlookers.

Within a year, thousands of Cherokee became literate and even started translating the King James Bible. A visitor reported seeing Cherokee letters written on or cut into fences, trees, and pieces of bark and boards lying all around. The novel alphabet had not been taught in school. The Cherokee had neither books nor paper. They had learned from each other.

A few years later, the Cherokee had a constitutional government and a newspaper in their own language and English, the *Cherokee Phoenix*.

What do you think of that?

How about the response of the Japanese to the total quality-control methods of mathematical statistician W. Edwards Deming? Born in a tarpaper shack in Iowa, Deming earned a scholarship to Yale and a Ph.D. in 1928 in mathematical physics. He then worked for the U.S. Agriculture Department and the Census Bureau until the War Department sent him to Japan in the late 1940s to help rebuild that war-spent land.

Who would have guessed how quickly and completely the Japanese would take to his ideas? By the late 1950s, a foreman could tune in the Japan Broadcasting Corporation to get a correspondence course in quality control. The typical Japanese bookstore today has a section devoted to quality. Statistical theory is taught in high school. "Made in Japan" signified poor quality when I was a boy; now it means superior quality.

The Japanese know how great a change they've wrought and who sparked it. Here's John A. Byrne's telling description from the January 28, 1991, issue of *Business Week*.

> How great is Deming's influence in Japan? On the walls in the main lobby of Toyota's headquarters in Tokyo, three portraits hang. There is one of the founder and one of the current chairman. But Deming's is the largest of all.

So Japan, perhaps as devoted to her own culture as any country on earth, responded to an outsider, learned new ways, and leapfrogged over the manufacturing technology of her victors.

The experiences of the Cherokee with language and of the Japanese with math-statistics have proved that tremendous cognitive changes can take place quickly.

Wherever you go and whatever you suggest, of course, you'll find plenty of doubters, people who want guarantees before they'll try anything new. And some of these people, I imagine, are scoffing right now at the two examples I've just given. They're dying to tell me that the Cherokee and the Japanese live in cohesive societies unlike ours. It's an obvious objection, and I have no guarantees we can do as well.

But then we Americans, diverse as we are, haven't exactly failed to pull together when we realized what was needed. I learned anti-war poetry and lived through the period of doubts before World War II, the America Firsters, Father Coughlin's diatribes, Lindbergh's praising the Nazis, and all that. Yet we went to war united.

I think any country could carry out a cognitive revolution, I believe the world could do it, if really required. It's part of my faith in human ingenuity.

Use a little imagination.

Think, for example, what would happen if mathsemantic sophistication became the mark of the new intellectual and the language buffs had to challenge the math buffs to see who would dominate the new field. There's a superbowl I'd like to see.

So I'm optimistic about everybody moving fast once they want to become mathsemantically sophisticated.

The more immediate question is how do we get started. I don't know what new Deming or Sequoyah will awaken us, or what medium he or she will employ.

It could be a cartoon that opens our eyes. Something like Walt Kelly's Pogo looking at a mountain of discarded junk and uttering, "We have met the enemy and he is us." True, Charles Schulz's Linus did tell Charlie Brown, "No problem is so big it can't be run away from," but surely he didn't mean literally forever, did he? I read it as gentle irony, in effect a chiding. Was I wrong?

Perhaps our inspiration will be a picture book, a sort of mixture of Wanda Gag's *Millions of Cats* and Watty Piper's *The Little Engine That Could*. The little engine worries whether it can solve the problems of the proliferating millions and billions, but then decides to try, and *succeeds*, all the while puffing, "**MATH**seMANtics, **MATH**seMANtics, **MATH**seMANtics, . . ."

More likely a music video will rock the world, something wilder than that part of Disney's *Fantasia* where Mickey Mouse plays the title role in Dukas's "The Sorcerer's Apprentice," and leave the message that it's up to us to halt the waves of consequences that we humans have set loose.

It could even be a full-length movie. Wolfgang Petersen's *The Never-Ending Story* gives us a literate and thoughtful child saving the world from "The Nothing" that had been created by adult apathy. Perhaps someone could make a prize-winning film about somebody saving the world from "The Overload" now being created by adult activity: *The Never-Ending Lines*; British title, *Queues Galore*.

I hope I'm around to enjoy the way some inspired artist makes sure that the message hits home. Meanwhile, some other practical questions are: For how many problems do we need mathsemantic ability? How fast are such problems accumulating? How quickly do we need to get started on our mathsemantic revolution? How much time do we have, really?

My wife keeps mentioning that she was born and has lived at the right time. She grew up in a New York so safe that she played hide-and-seek with her friends in a Greenwich Village subway station. She married and became a young full-time homemaker and mother who took our children to Central Park once or twice a day. No sooner had she met her responsibilities to our girls, seen them grown and gone from the nest, than she was invited to teach drama, her early love, part-time. In a few years she became full-time head of drama and chair of the fine arts department at a K–12 private girls school. She now looks forward calmly to the day when we might both choose to relax or whatever.

I don't think she fears the world is headed for sudden catastrophe or that it won't be worth surviving in. I think she just feels everything's getting more crowded, more precarious, more dependent on getting the details right, less forgiving of malfunction, less sized for and concerned with the individual, and therefore less rewarding. Konrad Lorenz called it "The Decay of the Humane."

I feel more like the boy apprenticed to Roger Bacon in the story about the philosopher's head. Bacon had fashioned through alchemy, as I retell the tale, a head of stone in the hope it would reveal secret knowledge. But he had exhausted himself and charged the boy to watch the head and come at once down the hall to wake him should anything happen.

Anon, in a lull during which the boy's mind could not help but wander, the stone head trembled slightly and whispered, "Time is coming." The boy was quickly alert and across the room. As his hand touched the door, he looked back. All was still and as before. "Did I doze off," he asked himself, "and dream?" He slipped back to scrutinize the head. "Nothing's changed," he told himself. "If I wake the Master now so shortly after he's gone to rest, and then have nothing to show, he'll be angry. Besides, if I heard aright, the voice said only, 'Time is coming,' so nothing has yet happened."

And as the boy hesitated, the head shook, gave off a puff of smoke, and said smartly, "Time Is Now."

The boy again sped across the room, determined not to repeat his initial error. But as he neared the door he pictured the long corridor down which he would need to travel to wake the Master, who would then need to travel the same route back. "Is there time," he asked, now filled with fear, "for the trip? The voice said, 'Time Is *Now*,' so there's no time left." The boy opened the door but looked back in despair. "If I leave, no one will hear the secret. What would my Master want me to do?"

And as the boy stood, one foot in the room, one in the corridor, hesitating, the stone head discharged a great black cloud, roared, "TIME IS PAST!" and shattered into a thousand pieces.

I've told that story many times to myself and others who hesitated. It usually gets action.

Another I use is Victor Hugo's "Fight with a Cannon."

A young naval officer fails to tie down a cannon, which later starts plunging erratically across the gundeck as the ship pitches in the waves, crushing the bulkheads and sides, thereby threatening to open the ship to the sea, sinking her and all aboard.

The officer takes an iron rod and goes below to fight a fabulous duel with the monstrous cannon, which rolls, hesitates, turns unpredictably, and hurtles back at him, as he dodges for his life. Finally, through great daring the young officer manages to jam the rod in the cannon's wheels so that it can be subdued. All are saved.

The captain assembles the ship's company, pins a medal on the officer for bravery, and for dereliction of duty, orders him shot.

At the office, when it seemed an employee might first begin to endanger us, perhaps through unwillingness to ask for help, I'd ask him or her to read the Hugo story and all the initiates would smile. In that way, we all came to understand what was meant when anyone reminded us in open meeting, "We don't want any heroes."

In spite of these examples, I don't see our problems in apocalyptic terms. I agree with my wife that things are just tending to get less comfortable. As Margaret Mead put it in 1970, "Life is going to get steadily worse but we're still going to be here."

Thus, late as it may be for us to gain the mathsemantic skills we need, I think there's time. When I'm late, I goad myself with "better late than later." "Better late than never" strikes me as offering the wrong choice. But that "never" is still possible.

Time has a way of sneaking up on you. When things start to go bad, it's usually later than you think. I lived through the end of a business once that looked healthy almost to the day it died. It's like you haven't crashed yet, the engines are still running, but the gas is almost gone. . . .

Eastern Airlines in November 1990 was given $135 million by the bankruptcy court to enable the ailing carrier to operate through March. It ceased operation in January. *Aviation Daily* reported that Eastern's trustee, Martin Shugrue, gave a speech in Washington on January 15, less than a week before all further flights were canceled, a tough assignment.

In the excerpt that follows I've changed the words "Eastern's people" to "All of us" to see how Shugrue's remarks would sound in a larger context. Imagine you're hearing the last speaker on a worldwide broadcast about whether our efforts to solve the world's problems will be in time.

> Whether we make it or not, our strides have been remarkable. All of us have reason to be proud. At the least we've demonstrated great spirit and indomitable resolve. At the most we're leaving a legacy of creativity, new ideas, and incredible ingenuity.

I'm being dramatic but not totally unreasonable. People have been around perhaps two and a half million years. The dinosaurs dominated the earth for one hundred sixty million years before they faded. That means we have about one hundred fifty-eight million more years to go to beat their record. If asked about our chances, shouldn't we answer modestly?

A childhood puzzler went, "If two fast-breeding bugs, which double their population every minute, are put in a bottle at eleven in the morning, and the bottle will be full at noon, when will it be half full?"

The follow-up question asked the next day was, "If the bottle was half full at eleven thirty, when would it be full?"

Think about it until you're sure.

In absolute numbers the recent increases in human population are unprecedented. That's a *sure* thing, beyond question to anyone who's willing to look and who comprehends large numbers, rates of growth, and simple statistical analysis.

That our unprecedented increases have already harmed the environment is *robustly sure* to all willing to trust the evidence of their own eyes and a flood of reports. Sensing the serious risk of our present trends then requires only a beginner's ability to follow forecasts.

Some people with an interest in this book have asked me to sketch briefly "what a mathsemantically literate society would be like ... to round out the book and underscore the practical ramifications." I'm having trouble with the suggestion. Somehow it seems inappropriate that I, who so strongly believe in self-determination, in multiplicities that can't be reduced to single instances, and in the impossibility of uttering the whole truth about anything, should attempt a summary forecast. My optimism depends not on having a plan, but on you and others finding hosts of avenues for development that escape me.

Perhaps the Outer Hebridean genes my father furnished steer me more toward what might happen if you and they don't find those developmental openings, if we continue the divorce between mathematics and semantics.

Proposition 29: Math without meaning could mean death by the numbers.

Such a dour proposition occurs naturally to one raised on the toast, "Here's to, who's with us, damn few, and they're all dead." My surname ancestors, like almost every people, more than once rallied to fight and lost. My line left for a new world. That option has since disappeared. I'd rather not lose again, for all our lines.

A mathsemantically sophisticated people should be able to address large-scale threats in a calm but *timely* manner. That's my general forecast of what could happen.

We've witnessed in our time a good deal of mathsemantically unsophisticated talk about where we're all headed and some daring confrontations, people at sea risking their lives to save dolphins, that sort of thing. It gets attention. I understand how people could care enough about this world to take such risks. Yet if more of us were mathsemantically sophisticated, I don't think they'd find it necessary. I'd feel safer in a world that didn't need heroes.

The fact that any given activity hasn't brought one to a sad end before now doesn't mean it's safe. If, for example, every auto accident were fatal, then every living driver could honestly say, "I've never had an accident." So just because we're still here doesn't mean that we—and our world—aren't threatened.

I agree with Margaret Mead that it doesn't help to forecast total disaster for the environment and our species. It's too all-or-nothing, too extreme. It's an attitude capable of triggering a bigger spree under the motto, "Enjoy it while you can."

The *Philadelphia Inquirer* of March 14, 1990, published a letter to its editor stating, "There's no such thing as 'too many people.'" I also find that too extreme a position. Just because the earth has so far supported an ever-increasing number of people doesn't mean there's no limit. We post occupancy limits in many vehicles and spaces to warn of danger. Overloaded elevators, for example, may be unable to rise, may slowly sink, or may plunge, perhaps to provide a smashingly ironic proof that numbers can be deadly.

Finding the safe carrying capacity of the earth is certainly much more complex than finding that of an elevator. But, because the earth carries all of us, its capacity is also considerably more important. Its ability to raise still more people to higher levels of consumption and well-being must end somewhere. "Where?" you ask. Well, we aren't sure even now whether in total we're rising, slipping, or sinking.

Today, that's the biggest mathsemantic question of all.

A

Quizzes

The instructions and questions used in the recruitment quiz for secretarial and clerical workers appear below, followed by some questions from the recruitment quiz for analysts and lawyers.

QUIZ FOR PROSPECTIVE EMPLOYEES

This test covers a number of different subjects and is intended to give some information on the basic skills and knowledge of all applicants. Therefore, no one would be expected to do well in all sections. Do as much as you can.

1. Solve the following problems in addition:

```
  1.32
 21.06    2.0
  1.07     .30              1 hr.  31 min.
  9.83     .9    2 apples   2 hrs. 50 min.   3 one-way trips
  5.26    106    5 oranges  6 hrs. 12 min.   2 round trips
 _____    ___    _____  _____    _____
                                             round trips
```

2. Solve the following problems in multiplication:

171	1246	1.407	16 travelers	-3
3	26	.32	2.5 hours	4.2

3. Solve the following problems in division:

$3)\overline{15}$ $3)\overline{15.1}$ $4.1)\overline{16.976}$ $30 \text{ min.})\overline{2 \text{ hrs. } 12 \text{ min.}}$

4. Round the following numbers to the nearest whole number:

63.1	271.9	3.5	.098	7 thousand
____	____	___	____	_____

5. Round the following numbers to the nearest two decimal places:

63.1	271.986	.0250	1.1049	3 million
____	_____	_____	_____	_____

6. Round the following numbers to the nearest hundred thousand:

486,440	3,751,120	.098	$-8,763,429,019.678$
_____	_____	____	_____

7. Express the following items as percentages to the nearest whole percent:

$\frac{7}{10} =$ ____ $.814 =$ ____ $12 \div 3 =$ ____ $\frac{3}{5} =$ ____ $.9 =$ ____

8. Solve the following problems:

$\frac{x}{2} = \frac{3}{6}$, $x =$ ____ $\frac{20}{x} = \frac{5}{2}$, $x =$ ____

9. What large American cities do you think are represented by the following station codes:

ATL _____	SFO _____
NYC _____	PHL _____
CHI _____	MIA _____
PHX _____	STL _____
DTT _____	DCA _____

10. What is the approximate population of the United States?

_____

11. About how far is it from Philadelphia to New York?

_____

12. About how far is it from Philadelphia to Los Angeles?

_____

13. To have the best chance of a letter's being received in San Francisco by July 15, you would mail it [from Philadelphia] on ( ) July 12, ( ) July 11, ( ) July 10, ( ) July 9, or ( ) July 8.

The following questions are from the recruitment quiz for analysts and lawyers.

14. Which is generally the most important in determining the reliability of a sample: ( ) the absolute size of the sample, or ( ) the percentage which the sample is of the entire population, or ( ) the absolute size of the entire population.

15. Would it affect the number of trips of 500 miles or more which would be made, if the local airport were 10 miles farther away from the prospective passengers? (Yes or No) _____
Why?_____

16. How good a fit does a coefficient of correlation of 1.0 indicate?

_____

17. In general practice, which is the most important to the reliability of a sample: _____size of sample; _____control of sampling bias; _____survey method (i.e., mail, telephone, personal interview, etc.); _____quality of editing; or _____tabulating method.

APPENDIX

Quiz index and answers

Each **bold-face numeral** below indicates a *page* where a discussion of that problem begins in the text. A dash indicates that the problem is not treated in the text. Nevertheless, answers to all quiz problems appear at the end of this appendix.

1. Solve the following problems in addition:

```
  1.32
 21.06    2.0
  1.07     .30              1 hr.    31 min.
  9.83     .9    2 apples   2 hrs.   50 min.   3 one-way trips
  5.26    106    5 oranges  6 hrs.   12 min.   2 round trips
 ─────    ───    ─────────  ──────────────    ────────────────
  176     176        4           18           19 round trips
```

2. Solve the following problems in multiplication:

```
  171       1246       1.407      16 travelers       -3
    3         26         .32       2.5 hours        4.2
 ─────    ───────     ──────    ──────────────    ─────
   55      55, 178       55           55            55
```

3. Solve the following problems in division:

$$\overset{-}{3)\,15} \qquad \overset{-}{3)\,15.1} \qquad \overset{-}{4.1)\,16.976} \qquad \overset{59}{30\,min.)\,2\,hrs.\,12\,min.}$$

4. Round the following numbers to the nearest whole number:

63.1	271.9	3.5	.098	7 thousand
143	**143**	**144**	**75, 145**	**144**

5. Round the following numbers to the nearest two decimal places:

63.1	271.986	.0250	1.1049	3 million
144	**144**	**144**	**144**	**145**

6. Round the following numbers to the nearest hundred thousand:

486,440	3,751,120	.098	−8,763,429,019.678
144	**144**	**145**	**145**

7. Express the following items as percentages to the nearest whole percent:

$$\frac{7}{10} - \underline{\textbf{343}} \qquad .014 - \underline{\textbf{244}} \qquad 12 \div 3 = \underline{\textbf{244}} \qquad \frac{3}{5} = \underline{\textbf{244}} \qquad .9 = \underline{\textbf{244}}$$

8. Solve the following problems:

$$\frac{x}{2} = \frac{3}{6}, \quad x - \underline{\textbf{82}} \qquad\qquad \frac{20}{x} = \frac{5}{2}, \quad x = \underline{\textbf{82}}$$

9. What large American cities do you think are represented by the following station codes:

ATL _____ — _____   SFO _____ — _____

NYC _____ — _____   PHL _____ — _____

CHI _____ — _____   MIA _____ — _____

PHX _____ — _____   STL _____ — _____

DTT _____ — _____   DCA _____ — _____

10. What is the approximate population of the United States? 148

11. About how far is it from Philadelphia to New York?   158

12. About how far is it from Philadelphia to Los Angeles?   160

13. To have the best chance of a letter's being received in San Francisco by July 15, you would mail it [from Philadelphia] on ( ) July 12, ( ) July 11, ( ) July 10, ( ) July 9, or ( ) July 8. 181

14. Which is generally the most important in determining the reliability of a sample: ( ) the absolute size of the sample, or ( ) the percentage which the sample is of the entire population, or ( ) the absolute size of the entire population.   257

15. Would it affect the <u>number</u> of trips of 500 miles or more which would be made, if the local airport were 10 miles farther away from the prospective passengers? (Yes or No)   206
    Why?   206

16. How good a fit does a coefficient of correlation of 1.0 indicate?
    _____259_____

17. In general practice, which is the most important to the reliability of a sample: _____size of sample; _____control of sampling bias; _____survey method (i.e., mail, telephone, personal interview, etc.); _____quality of editing; or _____tabulating method. 260

    The answers we have counted as correct are given below. The lowercase letters at left identify the separate problems within the numbered problem groups.

    1a. 38.54
    1b. 109.2, also 109.20

1c.   7 fruit, also seven pieces of fruit and the like

1d.   10 hrs. 33 min., also 10 hours 33 minutes, etc., but not just
        10   33, nor any other numbers (such as 9 hrs 93 min)

1e.   3 1/2, also 3.5

2a.   513

2b.   32,396, also 32396 (i.e., without the comma)

2c.   .45024, also 0.45024

2d.   40 traveler-hours, also 40.0 trvlr-hrs and the like

2e.   −12.6, but half off for (12.6), because parentheses are inconsis-
        tent with the style of the problem

3a.   5

3b.   5.0$\overline{33}$, also 5.03... or any other such answer showing a repeat-
        ing digit

3c.   4.14048781..., also 4.14+ and other correct answers that go at
        least that far and indicate with the main answer, not just down
        below, that there is a remainder

3d.   4.4, also 4 2/5, 4 12/30, etc.

4a.   63

4b.   272, also 272., but not 272.0

4c.   4, also 3 accompanied with a plausible argument for alternately
        rounding .5 up and down, but not otherwise

4d.   0

4e.   7,000, also 7000 (i.e., without the comma)

5a.   63.10, but not 63.1

5b.   271.99

5c.   .03, also 0.03 (even though not in style of question)

5d.   1.10

5e.   3,000,000.00, also 3000000.00 (i.e., without the commas)

6a.   500,000, also 500000 (i.e., without the comma)

6b.   3,800,000, also 3800000 (i.e., without the commas)

6c.   0

6d.   −8,763,400,000, also −8,763,400,000. (i.e., with useless decimal
        point), but half off for parentheses instead of minus sign

7a.   70%, half off if no percentage sign

7b.   81%, half off if no percentage sign

7c.   400%, half off if no percentage sign

7d.   60%, half off if no percentage sign

7e.   90%, half off if no percentage sign

8a.   1

8b.   8

9.   Atlanta, New York (City), Chicago, Phoenix, Detroit, San Francisco, Philadelphia, Miami, St. (Saint) Louis, Washington (D.C.), treated as ten questions in scoring, half off for each misspelling and/or failure to use initial capitals followed by lowercase letters

10.   260,000,000 would be about right in 1994, also anything from 200,000,000 (lower in past years) to 300,000,000, in numbers or words, with or without commas, but half off, for example, for 260 m or M, because the abbreviations could mean either thousand or million

11.   82 miles, also any answer from 70 miles through 100 miles, in numbers or words, with or without commas, but no credit if miles (mi., m, etc.) not stated

12.   2,389 miles, also any answer from 2,000 miles through 3,000 miles, in numbers or words, with or without commas, but nothing if miles not stated

13.   July 8

14.   the absolute size of the sample

15.   Yes, with any reasonable explanation (see chapter 19 for examples), but half off for yes with an unreasonable explanation

16.   perfect, also complete correlation, 100% fit, and the like

17.   control of sampling bias, but no credit for an equivocation if *any* part of it is incorrect

APPENDIX C

Mathsemantic propositions

Each proposition's chapter location is given in brackets.

1. Whenever we add *things*, we necessarily add *different* things, which we must then group under the same *name*. [1]

2. Although some people may qualify as experts, no one can say the whole truth about anything. [2]

3. For a count to make sense, you have to know what you're counting. [3]

4. Childhood semantics can impair adult mathsemantics. [4]

5. Some things that can be neither counted nor measured can still be savored. [4]

6. To improve a map, you have to adjust it to agree with the territory. [5]

7. To improve our mathsemantic maps, we must learn to think extensionally. [5]

8. In mathsemantic problems, it pays to tackle the semantics first. [6]

9. Math enthusiasts need to watch their language. [7]

10. Nothing is a mathsemantic problem. [8]

11. Without language, math is for the birds. [9]

12. Ordinary childhood semantics can't handle math. [9]

13. To make math your own, take vivid extensional lessons. [10]

14. Games can provide vivid extensional lessons. [10]

15. Until you work it out for yourself, two times two makes four only because the teacher says so. [11]

16. Mathematics is a limited and logically incomplete system of reasoning. [12]

17. Competent measurement requires facility in making and expressing approximations. [13]

18. Perfectionists can't measure up. [13]

19. How numbers start tells us less than their length. [14]

19A. A quantity's initial digits generally convey less information than its order of magnitude does. [14]

20. Physicists are natural mathsemanticists. [15]

21. Sloppy mathsemantic maps create real-world dangers. [16]

22. Maximizing takes mathsemantic initiative. [17]

23. Had we but world enough, and time,
    Bad mathsemantics were no crime. [18]

24. Collapsing multiple instances into a single one can be a particularly risky mathsemantic habit. [19]

25. To develop your own useful mathsemantic web, estimate checkable figures out loud. [20]

26. Beware of figures that, like projections and certified financial statements, seem to be meant as estimates of events but which specifically disclaim that responsibility. [21]

27. Percentages are dangerous social and economic tools that appear easy only to math teachers and the inexperienced. [22]

28. Understanding statistics is difficult yet vital; difficult because statistics bear so many mathsemantic burdens, and vital because statistics increasingly guide our lives. [23]

29. Math without meaning could mean death by the numbers. [24]

Selected sources

As this book is a "first," I can't direct you to other works that specifically recognize mathsemantics as such. Depending on your interest, however, you might find useful the sources I've listed and briefly described below (in two categories, books and organizations).

BOOKS

Bridgman, P. W. *The Logic of Modern Physics*. New York: Macmillan, 1927. Still the best introduction to that form of extensionalization known as operational definition.

Cohen, Patricia Cline. *A Calculating People: The Spread of Numeracy in Early America*. Chicago: University of Chicago Press, 1982. Insights into our mathsemantic history.

Feynman, Richard. *"Surely You're Joking, Mr. Feynman!": Adventures of a Curious Character*, and its sequel, *"What Do You Care What Other People Think?": Further Adventures of a Curious Character*. New York: Norton, 1985 and 1988. Paperback editions, New York: Bantam, 1986 and 1989. A scientist's romps that illustrate autonomy and mature mathsemantics.

Hayakawa, S. I. *Language in Thought and Action*. New York: Harcourt Brace Jovanovich, 1963 (or any later edition). The best selling book on general semantics, slanted for use in English courses.

Johnson, Wendell. *People in Quandaries: The Semantics of Personal Adjust-ment.* New York: Harper, 1946. General semantics slanted as the subtitle indicates.

Kamii, Constance Kazuko, with DeClark, Georgia. *Young Children Reinvent Arithmetic: Implications of Piaget's Theory.* New York: Teachers College Press, 1985. How to help, or thwart, children's grasp of math. See also its sequel, *Young Children Continue to Reinvent Arithmetic—2nd Grade,* 1989.

Korzybski, Alfred. *Science and Sanity: An Introduction to Non-Aristotelian Sys-tems and General Semantics.* 1933. The 5th edition (1993) is available from either the Institute of, or the Society for, General Semantics. The original and still surprising book on general semantics.

Lakoff, George. *Women, Fire, and Dangerous Things: What Categories Reveal about the Mind.* Chicago: University of Chicago Press, 1987. Why classical categories must give way to how we actually think.

MacNeal, Edward. *The Semantics of Air Passenger Transportation.* Norfolk, Va.: The Norfolk Port and Industrial Authority, 1981; distributed by the International Society for General Semantics. How to avoid bad mathsemantics in analyzing air passenger traffic.

MacNeal, Kenneth. *Truth in Accounting.* Philadelphia: University of Pennsyl-vania Press, 1939. Reprint. Houston: Scholars, 1970. My dad's book show-ing how accountants' figures don't tell what's really going on.

Newman, James R. *The World of Mathematics: A Small Library of the Literature of Mathematics from A'h-mosé the Scribe to Albert Einstein, Presented with Commentaries and Notes.* New York: Simon and Schuster, 1956. Four vol-umes of goodies, as advertised, that show the wide reach of mathematical effort.

Paulos, John Allen. *Innumeracy: Mathematical Illiteracy and Its Consequences.* New York: Hill and Wang, 1988. A teacher of mathematics tellingly de-plores its misuse.

Piaget, Jean. *The Language and Thought of the Child.* Translated by Marjorie Gabain. 2nd ed. London: Routledge & Kegan Paul, 1932. How children view language and thought. Also see other works of Piaget translated, printed, and reprinted at various times, such as *The Child's Conception of Number, The Child's Conception of Time,* and *The Child's Conception of the World,* each demonstrating that the natural semantics of children differs radically from that of most adults.

Pulaski, Mary Ann Spencer. *Understanding Piaget: An Introduction to Chil-dren's Cognitive Development.* Revised edition. New York: Harper and Row, 1980. A much easier and faster read than its sources.

Whorf, Benjamin Lee. *Language, Thought, and Reality: Selected Writings of Benjamin Lee Whorf.* Edited by John B. Carroll. Cambridge: MIT Press,

1956. How language influences both what one thinks is going on and how one chooses to deal with it.

ORGANIZATIONS

International Society for General Semantics (ISGS), P. O. Box 728, Concord, California 94522. Tel.: (510) 798-0311. Publishes *ETC: A Review of General Semantics*, a journal presenting a broad array of semantic interests.

Institute of General Semantics (IGS), 163 Engle Street, #4B, Englewood, New Jersey 07631. Tel.: (201) 568-0551. Publishes the *General Semantics Bulletin*, a yearbook tending to focus on central aspects of general semantics.

Jean Piaget Society (JPS), c/o Kurt Fischer, Harvard University, Human Development, Larsen Hall, Cambridge, Massachusetts 02138. Tel.: (617) 495-3614. Publishes *The Genetic Epistemologist* and holds annual symposia that address the development of children's cognitive abilities, including word and number languages.

American Statistical Association (ASA), 1429 Duke Street, Alexandria, Virginia 22314-3402. Tel.: (703) 684-1221. Publishes *Journal of the American Statistical Association* (JASA), *Chance*, and *The American Statistician*, which present and review the efforts of statisticians (both practical and theoretical) to represent various physical and social phenomena in numbers.

American Mathematical Society (AMS), P. O. Box 6248, Providence, Rhode Island 02940-6248. Tel.: (401) 455-4000. Publishes *Notices of the American Mathematical Society*, which often contains news of efforts to reach mathematics teachers and nonmathematicians, as well as more technical periodicals. About one member in three is also a member of MAA, the next listing.

Mathematical Association of America (MAA), 1529 Eighteenth Street, N.W., Washington, D.C. 20036. Tel.: (202) 387-5200. Publishes several periodicals, including *The American Mathematical Monthly*, which includes biographical sketches of mathematicians and invites discussions of collegiate mathematics and methods of instruction. About one member in four is also a member of AMS, the previous listing.

American Economic Association (AEA), 2014 Broadway, Suite 305, Nashville, Tennessee 37203. Tel.: (615) 322-2595. Publishes *American Economic Review* and other periodicals that present and review the efforts of economists to represent various decisional activities in numbers.

American Accounting Association (AAA), 5717 Bessie Drive, Sarasota, Florida 34233. Tel.: (813) 921-7747. Publishes *The Accounting Review*, books, and monographs that probe mathsemantic questions within the accounting framework.

American Bar Association (ABA), 750 North Lake Shore Drive, Chicago, Illinois 60611. Tel.: (312) 988-5522. Publishes *ABA Journal* and specialized journals, such as *Administrative Law and Regulatory Practice,* for its membership sections, which occasionally address the use of numbers and statistics.

General Index